ACID DEPOSITION

Commission of the European Communities

Acid Deposition

Proceedings of the CEC Workshop organized as part of the
Concerted Action "Physico-Chemical Behaviour of Atmospheric Pollutants",
held in Berlin, 9 September 1982

Edited by

S. BEILKE
Umweltbundesamt, Pilotstation Frankfurt, Federal Republic of Germany

and

A. J. ELSHOUT
N.V. Kema, Arnhem, The Netherlands

D. REIDEL PUBLISHING COMPANY

A MEMBER OF THE KLUWER ✱ ACADEMIC PUBLISHERS GROUP

DORDRECHT / BOSTON / LANCASTER

Library of Congress Cataloging in Publication Data

Main entry under title:

Acid deposition

     At head of title: Commission of the European Communities.
     1. Acid rain–Congresses. 2. Acid precipitation (Meteorology)–
Congresses. I. Beilke, S. (Siegfried), 1938–   . II. Elshout, A. J.,
1935–   . III. European Economic Community. IV. Commission
of the European Communities.
TD196.A25A25   1983   628.1'68   83-3235
ISBN 90-277-1588-2

The Workshop was organized by the
Commission of the European Communities,
Directorate-General Science, Research and Development, Brussels,
within the framework of the Concerted Action
"Physico-Chemical Behaviour of Atmospheric Pollutants"

Publication arrangements by
Commission of the European Communities
Directorate-General Information Market and Innovation, Luxembourg

EUR 8307
Copyright © 1983, ECSC, EEC, EAEC, Brussels and Luxembourg

LEGAL NOTICE
Neither the Commission of the European Communities nor any person acting on behalf of the
Commission is responsible for the use which might be made of the following information.

Published by D. Reidel Publishing Company
P.O. Box 17, 3300 AA Dordrecht, Holland

Sold and distributed in the U.S.A. and Canada
by Kluwer Boston Inc.,
190 Old Derby Street, Hingham, MA 02043, U.S.A.

In all other countries, sold and distributed
by Kluwer Academic Publishers Group,
P.O. Box 322, 3300 AH Dordrecht, Holland

Printed in The Netherlands

C O N T E N T S

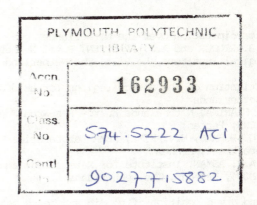

P R E F A C E

This volume contains the proceedings of a workshop on "Acid Deposition" held on September 9, 1982 at the Reichstag in Berlin.

The workshop was organized within the framework of project COST 61a bis* of the Commission of the European Communities (CEC), as a satellite meeting to the joint meeting of Working Parties 4 (Pollutant Cycles) and 5 (Transport and Modelling – Field Experiments) which was held on September 7 – 8 at the Federal Environmental Agency in Berlin.

The organisation of the scientific workshop programme was in the hands of a CEC Acid Deposition Task Force consisting of nine scientists actively participating in projects COST 61a or COST 61a bis : Beilke (Federal Republic of Germany), Brosset (Sweden), Elshout (The Netherlands), Gravenhorst (France), Liberti (Italy), Ott (CEC, Belgium), Penkett (United Kingdom), Sandroni (CEC, Italy) and Zephoris (France).

The workshop programme consisted of the presentation and discussion of five invited review papers covering different areas of the acid deposition problem and of ongoing work on topics which are relevant to acid deposition. Emphasis was placed on new scientific ideas and on indicating the most important areas of uncertainty, including the specification of a series of key research needs in the field of acid deposition. This was done out of concern for damage to forest and freshwater ecosystems in various European regions attributed at least in part to acid deposition originating from air pollution.

The discussion of research needs relevant to the acid deposition problem was mainly confined to physico-chemical processes occurring in the lower atmosphere.

Most of the research needs specified in these proceedings are contained in the paper entitled "Acid Deposition – The present situation in Europe" prepared by S. Beilke and discussed by the CEC Task Force and by the participants of the Berlin workshop. A specification of research needs on environmental effects was not attempted in this paper since such effects are not investigated within project COST 61a bis. Some research needs on ecological effects were specified by B. Ulrich in his paper on "Effects of acid deposition".

----

* COST   : Cooperation Scientifique et Technique
  61a   : Physico-chemical behaviour of SO_2 in the atmosphere
            (1972–1976)
  61a bis : Physico-chemical behaviour of atmospheric pollutants
            (1979–1983)

The research requirements outlined in these proceedings should provide a sound basis for defining future work within COST 61a bis and the possible follow-up project (1984-1985) in which more emphasis will be given to the acid deposition problem.

In my opinion, these proceedings give a fairly complete review of the research scene in this area, at least in the countries of the European Communities.

I am indebted to my colleages Dagmar Biehn, Christa Morawa, Helga Müller and Rolf Sartorius from the Federal Environmental Agency for their assistance in the technical organisation of this workshop. Finally, I would like to express my gratitude to Professor Georgii (University of Frankfurt, Federal Republic of Germany) and Professor Brosset (IVL, Gothenburg, Sweden) for acting as session chairmen.

Frankfurt,
December 1982   Siegfried Beilke

ADDRESS OF WELCOME

by

Prof. J. SCHMOELLING
Director of Division Air Quality Management; Noise Abatement,
Federal Environmental Agency

Ladies and Gentlemen,

On behalf of the President of the Federal Environmental Agency,
Freiherr von Lersner, I herewith would like to welcome you in Berlin in this impressive building!

I am particularly glad to welcome our friends, Mr. Ott, Mr. Versino and Mr. Stief-Tauch, from the Commission of the European Communities to whom the Reichstags-Building is already more familiar than to me!
This building is one of the few historical sites that survived in Berlin. It is called the German Reichstag, and Reichstag in this case means German Parliament.

The concept of the German Reichstag has been an integral part of Germany's political history since the Middle Ages (12th century).

The Reichstag developed from the representative assembly of the Franks into a diet comprising three chambers: for the electoral princes, the princes and the cities of the Reich. From 1663 on the Reichstag became a permanent congress of deputies with its seat in Regensburg, a political body with no constitutional contribution to the evolution of a homogeneous German nation. The dissolution of the ancient German Reich 1806 also meant the end of the Regensburg Reichstag.

After the foundation of the Reich 1871, its representative assembly was called the Reichstag which certainly stood for the unity of the Reich Bismarck strived for but the influence of the representatives on the affairs of government remained small, their powers of decision were also limited. The Weimar constitution of 1919 proclaimed the Reichstag, representing the population, to be the supreme bearer of state authority.

Today, the German Reichstag considers itself as the political successor of the Reichsstag of the Weimar Republic. The Basic Law of the Federal Republic of Germany has intentionally referred back to the tradition of the Reichstag of 1919 and the present German constitution preserves the spirit of liberal parliamentary democracy. This building here originally has been built in 1894 and since then was called the Reichstag - using the term for the former German parliament as a name for the building. It was meant to be its permanent home but due to the very particular political situation in our country it is today not the seat of the Federal Parliament.

The reason is that since Berlin is not the capital of the Federal Republic of Germany, the parliament is not permitted to get together in this city, only committee meetings may take place. Although it is a parliamentary building, however, non-governmental meetings are admitted in certain exceptional cases. And, the today's workshop is one of those exceptions and I am very grateful to the President of the German Bundestag who is the formal owner of this building for having granted us access to this impressive historical site.

Ladies and Gentlemen,

we did not meet today for talking history - the reason for this workshop is a more actual one! As all of you know, it is not our passion for nature and its life that makes us investigate the phenomenon which we call "acid rain" but rather the growing awareness that there is little doubt that acid precipitation has its significant share in damages to ecosystems. According to our present preliminary evaluations approximately seven percent of all forests in the Federal Republic of Germany are damaged - and this figure seems to be growing rapidly.

Already in the past the Federal Government has made significant efforts toward air quality management and is resolved to initiate strong actions to avoid further damages also in this area.

In this connexion I may call your attention to the last week's agreement of the Federal Ministers which clearly shows the seriousness of the problem as well as of the remedial measures to be taken:
new power plants may not emit more than 400 mg SO_2/m^3, in exceptional cases 650 mg/m^3. With respect to existing plants it was decided that they shall be subject to modernization within ten years in order to meet the same standards or have to be closed down within five years. These and other parallel measures let us expect a decrease in SO_2 emissions in our country from 4 million tons/a in the early seventies to presently 3,5 million tons/a, to finally 2 million tons/a in the nineties.

Ladies and Gentlemen,

Your host today is the Federal Environmental Agency. This Federal authority was established in order to give scientific advice to the Federal Government in environmental matters. In order to fulfill this mandate it is necessary - among others - to know atmospheric processes and to propose most efficient measures as well as to set priorities for air pollution control. One of the most useful tools in this respect is international co-operation in research into atmospheric processes.

It is one of the achievements of the Commission of the European Communities to have perceived this necessity in a very early stage and to have created the COST 61 a 1 and 2 Action Programme which forms an appropriate framework for co-operation and coordination.

But, not only scientists have to collaborate internationally - political planners and decision-makers have to talk together too in order to reach agreement about harmonized measures to prevent damages to our environment. It is a common place already that transport of air pollutants cross any border, and in particular acid precipitation is a transboundary and large-scale European problem.

In this connexion I should like to point to the results of the international conference on acid rain in Stockholm last June. There it was acknowledged that "with respect to air pollution it is time to say 'what goes up must come down', irrespective of the regions which will benefit from air pollution abatement. The development of scientific background as well as of abatement technology are expected to occur gradually and therefore do not justify delaying the use of available abatement technologies. There is a need to establish clear goals for reduction of emissions of sulphur dioxide. Stacks cannot replace measures at the source".

Against the background of those brief reflections I now would like to open the today's workshop on acid deposition and wish you every success for your discussion and conclusions!
Thank you!

A C I D   D E P O S I T I O N

THE PRESENT SITUATION IN EUROPE

S.Beilke

Umweltbundesamt,Pilotstation Frankfurt
Feldbergstrasse 45

1. INTRODUCTION

In the light of recent strong public interest in the "acid rain" issue,an attempt will be made to put together our present knowledge on acidic deposition to the earth surface. Since both wet and dry deposition contribute to the atmospheric input which may be responsible for a harmful impact on the environment,both processes will be considered here. Main emphasis will be given to the formation and removal of acidic substances under special consideration of the important contributions within projects COST 61a (1972-1976) and COST 61a bis (1979-1983) of the Commission of the European Communities. Although important contributions to the understanding of acid deposition were made through the OECD-LRTAP-project(1972-1976), this work is not dealt with since main emphasis was placed on investigating transport of SO_2 over long distances.

2. ENVIRONMENTAL EFFECTS

Interest in the acid deposition problem originates from a series of reports according to which acid rain is responsible for some environmental effects observed in certain European regions.These effects concern aquatic and terrestrial ecosystems as well as materials. However,an assessment of environmental impacts is difficult to make because of the complex nature of the interaction of acidic inputs with aquatic and terrestrial ecosystems (Hutchinson and Havas,1980).

As far as aquatic ecosystems are concerned,it is generally
accepted that a series of lakes and streams in southern
Scandinavia became increasingly acidic during the last two
decades(International Conference on the Effects of Acid
Precipitation,1976).
There is also general agreement among scientists that this
phenomenon is associated with the low buffering capacity in
and around lakes which make them particularly vulnerable to
acidification(International Conference on the Effects of Acid
Precipitation,1976).
There is also a broad consensus,although not unanimity,that
acid rain plays the major role for acidification of such
aquatic systems(Sandefjord Conference,1980).

Among the scientific community there is also a substantial
agreement that damage to terrestrial ecosystems(forests,
soils,crops) due to acid deposition is less well documented
than damage to aquatic ecosystems.

As far as terrestrial ecosystems are concerned,for a long
time most attention was paid to direct effects by gaseous
pollutants near emission sources.Recently,however,interest
has concentrated on regional effects at sites remote from
emission sources.Of special interest in some European countries is the possible environmental impact of acid deposition
on forest ecosystems.It is evident that coniferous forest
stands in different European areas show more or less severe
symptoms of illness the reasons of which are complex and are
traced back to biotic and/or abiotic stress factors.
Effects on forestry are mainly observed in some countries of
central Europe(Poland,Czechoslovakia,Austria,Switzerland,
German Democratic Republic,Federal Republic of Germany)
whereas in Scandinavia no such effects have been found despite intensive studies during the SNSF-project.
In this region the main concern is concentrated on freshwater
ecology and its response to atmospheric inputs by acidic rain
or snow containing substances transported long distances.
Among the different factors which are thought to be responsible for the effects on forest stands are acidic inputs due
to dry and wet deposition.Such acidic inputs may act either
directly or indirectly by affecting the soil.
According to Ulrich(1983),there is convincing evidence that
acidic atmospheric inputs increase soil acidity thereby
altering soil processes and leading to the damage of forest
ecosystems.On the other hand,Rehfuess(1981) concluded that
the damaging effect of acid deposition on forest soils is
not yet documented so far.
It should be noted here that in European terms both freshwater ecology and forestry ecology are areas of comparable
concern whereas in global terms the most important area of
concern has been and still is freshwater ecology.

Another potentially important impact of acidic deposition is
the effects it may have on the detoriation of materials
(for example monuments).
Although this problem is of importance in almost all pollu-

ted European regions,it is considered particularly important
in some countries of southern Europe such as Italy and Greece
(Liberti,1982;Sandroni,1982).

It should be emphasised that all harmful impacts on the environment should be seen in the wider context of total air
pollution including heavy metals,organics,oxidants ect. and
not only of acidic substances(Arndt et al.,1982;Prinz,1982).

3. ATMOSPHERIC ACIDITY

Acidity is the capability of a single substance or a system
of substances to produce protons by reacting with water
(Brønsted acids).
The acidity of an atmospheric system is determined by its
ability to transfer protons directly to or to cause the formation of protons within a receptor system.

Acidity is present in fog-and raindroplets,snowflakes and
aerosol particles mainly as H^+ and NH_4^+ and reaches the ground
with these bulk H^+-carriers.
In addition,weak and strong acids can be deposited directly
onto the different surfaces or they are formed on vegetation,
soil-and water surfaces when atmospheric gases are absorbed.
All of these processes may contribute to the acidification of
various ecosystems and are collectively called acid deposition
(FIGURE 1).

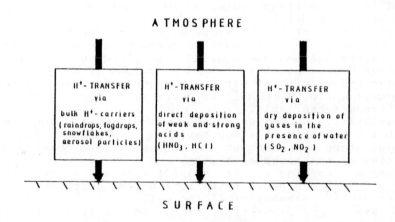

The free acidity of rainwater results from the interaction of
strong and weak acids with corresponding bases.
As a measure of the free acidity of rainwater very often its
pH is used.PH is the negative logarithm of the hydrogen ion
(proton) activity.The free acidity is not to be confused with
the total acidity of rainwater which is a measure for total
acidic wet deposition.
Atmospheric acidity is difficult to determine due to the

large number of substances which interact in a very complex way.
The acidity of an atmospheric droplet system[1] may be determined in terms of an ionic balance of cations and anions.
The most frequent cations of precipitation water in central Europe are H^+ and NH_4^+ and to a lesser extent Ca^{++}, Na^+, K^+ and Mg^+ which are balanced by the anions $SO_4^=$, NO_3^-, and Cl^-.
The relative proportions of the different anions seem to be relatively well known. Measurements in different parts of the Federal Republic of Germany show a majority of equivalent concentrations for sulfate(60-70 %) followed by nitrate and chloride(30-40 %) (Gravenhorst et al.,1980b;Müller et al., 1981).
Since in rainwater the dissolved ions are no longer directly associated with their original counter ions,it is only possible to reconstruct by indirect means the former composition of the parent compounds of the rain water constituents.
It is therefore difficult to identify quantitatively the acidifying agents.
In precipitation samples of the industrialized European areas, the sum of the concentrations of protons,ammonium-,sodium-, and calcium ions is generally counterbalanced by the sum of sulfate-,nitrate-,and chloride ions.It is reasonable to assume that sulfate,nitrate and a fraction of chloride are formed together with protons.The resulting acid systems are afterwards partially neutralized by dissolution of basic particles and gaseous ammonia i.e. the total acidity of the system increases due to neutralisation processes.
Measured pH-values therefore represent the free protons present in the analyzed system only and not the total amount of (Brønsted) acids.
The free acidity of rainwater is often measured by pH-electrodes.In Europe in many cases pH measures the strong acids present since the thus determined free H^+-concentration is to a large degree counterbalanced by anions of strong acids.
The potential acidity of various compounds(for example that of the weak cation acid NH_4^+) is,however,not seen by the pH-electrode(Tyree,1981).
The possibility can not be ruled out that acids which are weak in the pH-range of atmospheric rain may become important proton-donators depending on the biological and chemical properties of the receptor system.
For example,according to Ulrich(1983) weak acids like the cation acid NH_4^+ and heavy metal ions become important proton donators at the pH-range of living cells.
In the case of dry deposition,gaseous SO_2 is likely to act as the most important proton-donator to ecosystems in several areas of central Europe.

---

(1): Atmospheric acidity may also be determined using the approach of thermodynamic phase equilibria within an atmospheric system consisting of a liquid phase and a a gas phase containing the main elements of the acidity determining matrix
(Brosset,1979;Brosset,1983).

Acid deposition is a phenomenon which does not occur to the same extent in all European areas.It strongly depends on the amount and kind of emissions and on the climatological situation encountered and is therefore highly variable.

The natural acidity of precipitation and its pH is variable. Pure distilled water in equilibrium with the 335 ppm of atmospheric CO_2 results in a pH of about 5.6.
However,due to the incorporation of condensation nuclei and other aerosol particles of natural origin(soil,dust,sea spray) and due to the absorption of naturally occurring gases, the pH of a natural rain can be higher or lower than this value (Barrie et al.,1974;Charlson and Rodhe,1982;Seqeira, 1982;Delmas and Gravenhorst,1983).
Therefore,a pH of 5.6 cannot be used as a reasonable refe- rence value for defining an unpolluted or natural background acidity.
This is supported by measurements of pH and strong acids in some remote unpolluted marine and continental areas within COST 61a bis and other projects showing a wide range of pH values and strong acid concentrations(Gravenhorst et al., 1980a;Miller and Yoshinaga,1981;Delmas and Gravenhorst,1983). It is therefore difficult to assess exactly man's contribu- tion to the high acidity of rain water observed in some areas.

Nevertheless there is a broad consensus that central Europe is receiving acid wet deposition at rates in excess of that expected for unpolluted areas.It is also recognized that rain in air masses which have passed over industrial regions as far as hundreds of kilometers away is normally more acidic and contains higher concentrations of sulfate and nitrate than rain from other sections (KEMA Report,1981).

On a regional scale in central Europe,yearly volume-averaged values of acidity in precipitation range between 4.1 and 4.5.

Figures 2 and 3 show annual mean concentrations of sulfate (FIGURE 2) and nitrate(FIGURE 3) in rain (Wallen,1980) averaged for 1972-1976 on the basis of monthly means of 25 stations of the WMO-Background Air Pollution Monitoring Network(BAPMON).
The mean sulfate-background level gradually increases from less than 1 mg/l to ca. 5 mg/l in the maximum area in central Europe extending from south of England over northern France, Germany,Poland to part of the Soviet Union.
In the maximum area nitrate values up to 3 mg/l are found compared to less than 0.5 mg/l in the minimum area.

In the past,however,the pH value of bulk precipitation and not of rain samples was most often determined.Bulk deposi- tion samples represent the combined effect of total wet and particulate dry deposition as well as an undefined amount of gas phase dry deposition.The bulk deposition samples are usually more alkaline than wet-only samples because of the alkaline soil particles being deposited during dry periods (Winkler,1977).These field measurements should therefore be be interpreted with caution.

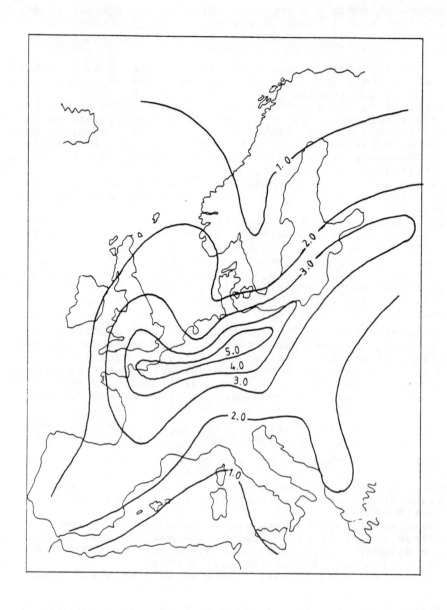

FIGURE 2 : Annual mean concentrations of sulfate in rainwater after Wallen(1980) for 1972-1976 on the
basis of monthly means of 25 stations of the
WMO-Background Air Pollution Monitoring Network
(BAPMON).

Sulfate concentrations are given in mg/liter.

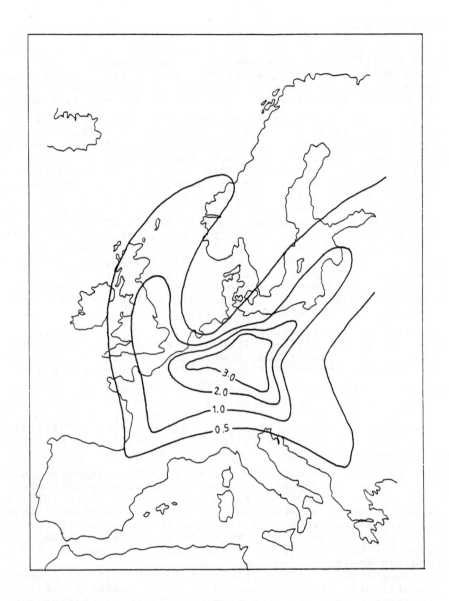

FIGURE 3 : Annual mean concentrations of nitrate in rainwater after Wallen(1980) for 1972-1976 on the
basis of monthly means of 25 stations of the
WMO-Background Air Pollution Monitoring Network
(BAPMON).

Nitrate concentrations are given im mg/liter.

4. EMISSIONS OF SO_2 AND NO_x

The anions sulfate,nitrate and chloride in rainwater can be
formed by different processes.It is now recognized that
gaseous sulfur and nitrogen oxides are the principal precursors for the anions sulfate and nitrate,respectively.
These oxides originate from both natural and man made sources.

On a global scale,the relative contributions of natural and
man made sources to the global production of sulfur and nitrogen oxides still remains uncertain in spite of considerable
progress achieved within project COST 61a bis.
Recent investigations within this project have shown that on
a global scale natural emissions contribute less than 40 %
to the total atmospheric emissions of sulfur
(Janssen-Schmidt et al.,1981).
Natural sulfur gas emissions consist mainly of volcanic
emissions of SO_2 (Jaeschke,1980;Sabroux,1980;Carbonelle,1981)
and biogenic sulfur emissions of $(CH_3)_2S$ from the ocean
(Nguyen et al.,1980) and reduced sulfur compounds such as
H_2S and CS_2 from tropical soils(Delmas et al.,1980).
An attempt to estimate the contribution of various natural
S-compounds to the global atmospheric sulfur cycle was made
by Varhelyi and Gravenhorst(1981).

For nitrogen oxides,natural sources such as lightning,bioge-
nic processes in soil,oxidation of ammonia etc.contribute to
ca. 50 % to the global source strenght of these acid forming
gases(Janssen-Schmidt et al.,1981;Böttger et al.,1978).

Most of the anthropogenic emissions of SO_2 and NO_x occur in
the industrialized regions of the northern hemisphere covering
less than 5 % of the earth surface(Europe,northeastern
America,China,Japan).According to Cullis and Hirschler(1980),
over 90 % of the man made sulfur is emitted in the northern
hemisphere.

On the regional scale of the industrialized regions of Europe,
man-made emissions of both sulfur and nitrogen oxides are
overwhelmingly dominant.An exception is Italy where volcanoes
provide the major SO_2-emissions.These volcanic sources inject
their sulfur compounds into atmospheric layers where they do
not directly influence the immediate vicinity of man's
environment but contribute to the atmospheric burden on a
much wider scale.
Man-made emissions of SO_2 originate mainly from the combustion
of fossil fuels as a consequence of their sulfur content and
to a lesser extent from smelting of sulfide metal ores and
other industrial processes.
Man made NO_x originates from the oxidation of atmospheric
nitrogen during combustion as well as from trace nitrogen
compounds in fossil fuels.Both stationary and mobile sources
contribute to the emission of NO_x.

---

(1):NO_x is the sum of NO and NO_2.Because of the rapid photochemical equilibrium during daytime both NO and NO_2 are
   treated as NO_x.

Considerable environmental concern was raised that via nitrification and denitrification of chemical fertilizers N_2O
could among others contribute to the acidification of atmospheric deposition after being converted to NO_x.
However,the total global N_2O-source strenght is only partly
caused by anthropogenic activities due to combustion processes
and fertilisation(Weiss,1981).
The overwhelming part of N_2O is photolysed to N_2.A few percent of the total N_2O-source strength(ca. 10%) is converted
by the reaction with singlet atomic oxygen to nitric oxide
in the stratosphere and subsequently form nitric acid.
Anthropogenic N_2O contributes on the order of 1 % - 3 % to
the total anthropogenic NO_x-source strenght as estimated by
Böttger et al.(1978) and can therefore be neglected in this
context.

Man-made emissions of SO_2 and NO_x in northwest Europe have
risen substantially during industrial times.
Estimates of Bettleheim and Littler(1979) show an increase of
SO_2-emissions by a factor of ten between 1860 and 1970 for
Europe excluding the USSR.
According to estimates of Fjeld(1976) and of Bettleheim and
Littler(1979),SO_2-emissions in Europe(without USSR) have
nearly doubled between 1950 - 1970.
However,as indicated in table 1,there are considerable dis-
crepancies between these authors about the absolute figures.

AUTHORS	1950 (tonnes SO_2/year)	1970 (tonnes SO_2/year)
Fjeld(1976)	24 x 10^6	49 x 10^6
Bettleheim and Littler(1979)	18 x 10^6	33 x 10^6

TAB.1 : Increase of SO_2-emissions in Europe(without U.S.S.R.)
between 1950 - 1970 according to literature release
data.

As these SO_2-emission estimates are mainly based on fossil
fuel consumption and sulfur content,they may be subject to
considerable error.
For example,in the Federal Republic of Germany,the
SO_2-emission in 1978 was nearly the same as in 1966 in spite
of an increase of the primary energy consumption by ca. 50 %
(Umweltbundesamt,1981).
In contrast to SO_2,NO_x-emissions have increased during this
period in the Federal Republic of Germany by ca. 50 %
(Umweltbundesamt,1981).
In the light of the substantial increase of man-made emissions of SO_2 and NO_x in north-west Europe since the onset of
industrialisation,it is reasonable to assume that both the

sulfate and nitrate content of precipitation should have
shown an increase since that time(KEMA report,1981).

5. TRANSFORMATION PROCESSES

After their emission,the gases SO_2 and NO_x undergo a variety
of physical and chemical processes(transport and conversion)
within the atmosphere until they or their reaction products
are removed from this reservoir.

As far as their chemical transformation is concerned,oxida-
tion of SO_2 and NO_x can occur by homogeneous gas phase
reactions,in aqueous droplets or water films and on atmospheric surfaces.All of these processes occur simultaneously.
The rates at which these gases are oxidized depend on the
specific environment considered and are highly variable.For
the concerned regions of central and northwestern Europe,the
oxidation by all of these mechanisms has to be taken into
account.

It is now accepted that the most important pathway for the
homogeneous gas phase oxidation,if not the most important
atmospheric oxidation route of all,is the reaction of SO_2
and NO_x with OH radicals which are formed due to photochemical activity(Cox,1976;Eggleton and Cox,1978;Calvert et al.,
1978;Schurath,1980).
The final product of the homogeneous SO_2-gas phase oxidation
are particles in the nucleation mode($r < 0.1 \mu$ m) consisting
of mixed aggregates of H_2SO_4-and H_2O-molecules.
These nuclei grow by coagulation into the accumulation mode
(r: 0.1 - 1 μm; Boulaud et al.,1978).
These fine particles smaller than 1 μm radius are quite aci-
dic(Junge and Scheich,1971;Gravenhorst,1978).Their chemical
composition is determined to a large extent by the reaction
products of the gases from which they are formed.
The chemical composition of the soil-derived coarse particles
($r > 1 \mu$m) differs considerably from the smaller particles.
The nitric acid products of the gas phase NO_2-oxidation

$$NO_2 \quad + \quad OH \xrightarrow{\quad M \quad} HNO_3$$

has a higher equilibrium vapor pressure than H_2SO_4 and is
therefore expected to not always form new particles in the
nucleation mode but rather to become attached to larger more
alkaline particles(Gravenhorst et al.,1979).

Since fine particles in the size range of 0.1 - 1 μm act
preferrentially as condensation nuclei for the formation of
cloud-and fog droplets,the acidity of such droplets and hence
that of falling rain can be determined,to a certain extent,
by the homogeneous gas phase oxidation products of SO_2 and
NO_x(Gravenhorst et al.,1980a;Fowler,1980).

The acidity in atmospheric cloud-and fog droplets can be
increased by an absorption and subsequent oxidation of SO_2

and NO_x within the aqueous droplet phase.These oxidation
processes are less well understood,particularly those involving NO_x.
Laboratory experiments have shown that the oxidation of SO_2
in bulk and dispersed aqueous systems is increased by catalysts such as manganese and iron(Barrie and Georgii,1976) and
by strongly oxidizing agents such as hydrogen peroxide(H_2O_2)
and ozone(Penkett et al.,1979).
However,the importance of these processes in the atmosphere
still remains unclear and reasonable conversion rates cannot
be given yet for atmospheric conditions.

Sulfuric acid and nitric acid are most likely the main constituents formed by these oxidation processes.
Absorption of ammonia and other alkaline substances may
follow so that protons are transferred to weaker acids or
form water again.

Due to the atmospheric transformation of SO_2 and NO_x to sulfate and nitrate during transport in air,the ratio between
these gases and their oxidation products will decraese with
distance from the source of these gases.

An extensive review of the different processes leading to
atmospheric acidity including a specification of the major
areas of uncertainty was made by Cox and Penkett(1983) in
these proceedings.

6. REMOVAL OF ACIDIC SUBSTANCES

Sulfur-and nitrogen oxides and their oxidation products are
removed from the atmosphere by two types of processes:
dry and wet deposition.

Dry deposition is the direct transfer of gases,aerosol particles and fog droplets from the atmosphere to surfaces
(water,snow,soil,vegetation) whereas wet deposition stands
for the transfer of substances via rain droplets.
Under conditions of high relative humidity salt particles
may deliquesce and become small droplets which may impact
on surfaces with greater efficiency than dry particles
(Chamberlain,1975).In such a case a distinction between dry
and wet deposition becomes blurred.
The atmospheric burden of sulfur-and nitrogen oxides is reduced by these deposition processes.At the same time the
surfaces on which they are deposited are subjected to chemical changes one of which is certainly an acidification if
the system has a low buffering capacity that means the system is not able to transfer protons to OH^- - ions.

There is considerable uncertainty about the relative proportions of acid contributed by dry and wet deposition.
This uncertainty is mainly due to a lack of knowledge about
dry deposition processes which makes estimates of this removal mechanism much less accurate than estimates for wet
deposition acidity.

6.1. Removal by dry deposition

Most measurements of dry deposition have been made inter-
mittently in selected areas of central Europe.The range of
deposition velocities(1) derived from these intermittend
measurements is between ca. 0.1 cm/s to 2 cm/s for SO_2 depending on a series of meteorological and surface parameters
(stability of the surface near atmosphere,kind of surface,pH
of soil and water surfaces,vegetation cover ect.).

For NO_x fewer dry deposition measurements are available.They
suggest that this process is not an effective sink for atmos-
pheric NO_x(Judeikis and Wren,1978;Böttger et al.,1978;
Beilke and Gravenhorst,1979).
There is some evidence that soils may be a source of NO
rather than a sink(Beilke and Gravenhorst,1979).
Dry deposition of NO_x occurs most likely in the form of the
gaseous NO-oxidation products NO_2,HNO_3,and N_2O_5.
As far as NO_2 is concerned,the range of deposition velocities
reported in the literature is between 0.01 - 0.8 cm/s
(Judeikis and Wren,1978;Böttger et al.,1978).
There is growing evidence that dry deposition of HNO_3-gas may
contribute appreciably to total removal of NO_x from the at-
mosphere(Fowler,1980).
Although no measured deposition velocities in the field are
available for HNO_3-gas,its high chemical reactivity implies
similiar,if not higher deposition velocities than for SO_2.
Sulfur and nitrogen compounds can be removed by dry deposition not only in gaseous form but also in particulate matter.
In addition to the parameters affecting dry deposition of
gases,deposition of particles is also dependent on particle
size.For the particle size range of the accumulation mode
where most sulfate and nitrate is found(0.1 - 1 μm particle
diameter) deposition velocities have a minimum and are not
expected to exceed 0.1 cm/s.
Some authors suggest that the low rates make dry deposition
of these particles only a small fraction of total dry deposition of sulfur and nitrogen(Fowler,1980;Garland,1977;
ISSA-Workshop,1977).
Recent investigations,however,seem to show that dry deposition of particles and gases can contribute equally to the
total deposition of these compounds in less polluted areas
(Koch,1979).
However,there is not complete agreement that particle deposition velocities are so low.For instance,in a field study
of the deposition of atmospheric particles onto a forest
canopy,deposition velocities in the order of 1 cm/s were
derived for relevant particle sizes and constituents
(Höfken et al.,1981).

---

(1): Deposition velocity is defined as the flux of a compound
     to the surface divided by the concentration of this
     compound at a reference level (normally two meters
     above the ground).

6.2. Removal by wet deposition

Wet deposition of sulfur and nitrogen compounds may be determined from measured composition and amount of rainfall.
An exact quantification of the different mechanisms responsible for the overall composition of rainwater is,however,
not possible due to the complex nature of processes by which
gases and particles find their way into raindrops.
Sulfur and nitrogen compounds are captured by cloud-and rain
drops by different mechanisms such as incorporation of cloud
condensation nuclei(CCN),impaction or Brownian diffusion of
particles or by dissolution of SO_2,NO_2,and HNO_3 and subsequent oxidation or dissociation.
Model calculations and measurements have shown that physico-
chemical processes within clouds(1) such as the incorporation of cloud condensation nuclei(CCN) and the oxidation of
sulfur IV species are the largest contributors to sulfate
content in rainwater and hence likely to its acidity
(Beilke and Georgii,1968;Fowler,1980;Gravenhorst et al.,1980).
Since the chemistry of aqueous NO-and NO_2-oxidation is only
poorly understood,the role which this reaction path may play
for nitrate formation and acidification of cloud water is
not known.
In the case of falling rain(2),the time in which the drops
remain airborne is too short for a significant oxidation of
SO_2 and NO_2(Beilke and Gravenhorst,1976;Fowler,1980;Penkett
et al.,1979).
However,the incorporation of aerosol particles by falling
raindrops may contribute to an extent to the sulfate-and
nitrate content and perhaps to the acidity of rainwater which
is not negligible(Meurrens and Lenelle,1976;Fowler,1980).

In spite of the considerable uncertainty about the relative
proportions,a comparision of the contributions of wet and
dry acid deposition should reveal a general tendency in such
a manner that dry deposition is relatively more important
for polluted regions than for unpolluted ones.This is due to
the fact that dry deposition is proportional to the acid
concentration in the near-surface atmosphere whereas,in large
areas in central and northern Europe,acid deposition by rain
depends mainly on the amount of rainfall(Georgii and Perseke,
1979).
If fog-and cloud droplets are important carriers of acidity
to ecosystems,than the spatial and temporal distribution of
the interaction of these hydrometeors with the earth surface
affects in addition the overall acid deposition.
A first attempt to follow the fate of SO_2 in a reaction
kinetic model and to compare homogeneous gas phase oxidation

---

(1): Physico-chemical processes within clouds are ofter
     called "rainout" or "in cloud scavenging"

(2): Physico-chemical processes transferring material to
     falling raindrops below cloud base are often called
     "Washout" or "below cloud scavenging".

wet chemical transformation and dry deposition was made by
Gravenhorst et al.(1978).

Summarizing our current knowledge,of the large number of factors which are thought to be of importance for the acid deposition problem,only relatively few are well known.
However,the causative chain leading to the environmental
effects observed in some European regions and the possible
contribution of acid deposition can most likely not be followed step by step due to the complex nature of this problem.

Instead,further research should be concentrated on the investigation and quantification of a series of key processes.

7. FUTURE RESEARCH NEEDS

The source-receptor relationship between pollutant emissions
and acid deposition in different European areas cannot be
quantitatively determined yet.

With the increasing recognition of the acid deposition problem
the desire has been expressed to limit and decrease the
acidity of precipitation and deposition.

However,if this is done effectively and at minimum costs,we
have to quantify at least some of the physico-chemical key
processes in order to come to a realistic description of the
source-receptor relationship in acid deposition.

In the following section the most important uncertainties are
indicated and a series of key research needs specified.
They have mainly resulted from discussions within the Task
Force "Acid Deposition" of project COST 61a bis.
Some use was also made of the conclusions of two workshops
(Alta,USA,1978;La Jolla,USA,1982) published in the corres-
ponding workshop-proceedings.

Research needs on ecological effects are not outlined since
such effects were not investigated within project COST 61a
bis.

7.1. Investigation of acidity trends

An important question in the acid deposition problem is
whether,and to what extent,total deposition has increased
during the last three decades in various European regions.
An answer to this question is not easy on the basis of the
available measurements for dry and wet deposition.
On the one hand,trend measurements for acid dry deposition
over the last decades are virtually non-existent.
On the other hand,the question of what the history of total
rainwater acidity has been during the last decades still
remains open,in spite of a number of discussions of this
topic during the last decade(Granat,1972;Kallend,1983;
Kallend et al.,1982;Winkler,1983).

A knowledge of past trends could help to answer the question

of whether,and to what extent,the recognized increase of SO2
and NO$_x$-emissions in Europe over the last decades
(Bettleheim and Littler,1979;Fjeld,1976) has affected the
acidity of rainwater.
Intercomparisions over large European areas for long time
periods require a consistancy of methology that is not available from historic measurements in Europe.This limits the
conclusions which can be drawn on trends in precipitation
composition and acidity.
Previous publications showing an increase of the geographical
extent and amount of free protons in precipitation between
the 1950's and 1970's in Europe should therefore be viewed
with caution.
This concern is shared by a recent publication within COST
61a bis presenting a statistical reevaluation of the precipitation acidity data from the European Chemistry Network
for the time period 1956-1976(Kallend,1983;Kallend et al.,
1982).
With respect to acidity trends,the following research needs
have been specified:

- High priority should be given to a rigorous analysis
  of existing precipitation chemistry data for trends
  and variability

- A rigorous analysis for trends of gases and aerosol
  composition is also necessary

- There is a need for a more rigorous analysis of
  existing European emission data for SO2 and NO$_x$ for
  trends and variability on a grid as fine as possible.

Another possible means of drawing conclusions about total
acid deposition trends can be the mass continuity approach:
"What goes up must come down".
On a global scale,this continuity approach provides a linear
relationship between total sulfur and nitrogen emissions and
total deposition of all the different forms of these species.
This approach is still acceptable on a scale of ca. 4000 x
4000 km if we look into the mean residence times of the main
compounds which determine atmospheric acidity.
However,such an approach is not valid and thus not helpful
to establish a source-receptor relationship on the scale of
a single European country.Specific areas within Europe might
experience significantly higher or lower acidic deposition
from a change in emissions.

7.2. Investigations of clouds

An important research need is the investigation of the physico-chemical processes in clouds otherwise no advance in understanding wet deposition as a major sink of acidic substan-
ces("in cloud scavenging") will be made.
Of particular interest is to

- measure pH,conductivity and chemical composition of
  cloud water in various types of clouds in different
  regions and seasons in both non-precipitating and
  precipitating clouds,fog,haze ect.(field measurements).In Europe two groups(Kallend et al.,1981;
  Elshout et al.,1983) have successfully started such
  measurements.

- measure simultaneously the chemical composition of
  different particle size ranges under special consideration of the cloud condensation nuclei(CCN) and
  ambient concentrations of the main acidity-determining
  substances(SO_2,NO_x,HNO_3,HCl,NH_3,O_3,H_2O_2,organic acids,
  alkaline and acidic particles) in drop-free air within
  and outside clouds(field measurements).
  The results of a series of measurements to study the
  chemical liquid water composition,the interstitial
  gaseous-and aerosol components within non-precipitating stratiform clouds and the surrounding clear air
  was recently reported by Daum et al.(1982).

- investigate the role which the ice-phase plays for
  the formation of acid precipitation(field-and laboratory measurements)

- develop local and regional models(parallel to such
  measurements in clouds)describing the efficiencies
  of clouds to incorporate the various substances.Such
  models should include dynamic processes such as ver-
  tical-and horizontal motion,droplet formation-and
  evaporation ect.as well as chemical conversion processes on a basis of a parametrization of heterogeneous chemical reactions(modelling).

Besides such investigations in clouds there is also a need
for determining realistic scavenging coefficients for the
various compounds below cloud base("Below cloud scavenging").

7.3. Transformation studies

An important problem is the investigation of chemical reactions
in the gas-and droplet phase.
There is a need for laboratory and field studies including all
the chemical species relevant to the acidification in the atmosphere.
Of special interest are the following research needs:

- A detailed knowledge of the chemical conversion of
  SO_2 and NO_x both in the gas-and liquid phase should
  be provided.Information on reaction rates,intermediate
  species leading to sulfate and nitrate is lacking.
  Of special interest is to investigate the mechanisms
  of the reactions of OH with SO_2 and NO_x and their
  effects on the HO_x budget on a regional scale.
  Such investigations are important to determine the
  rate-limiting factors in the formation of atmospheric

acidity.For example,substantial further research is
necessary to determine the rate-limiting steps in the
conversion of sulfur dioxide to sulfuric acid both in
the gas-and droplet phase under atmospheric condi-
tions.There is some recent evidence that the limiting
factors in the formation of sulfuric acid and of
acidity in a polluted atmosphere appears to be the
availibility of oxidizing agents such as OH,HO_2 radicals or H_2O_2(Rodhe et al.,1981).
As these oxidizing agents are directly linked to the
atmospheric NO_x-NMHC-cycle,the chemistry of these
compounds plays an important role for SO_2-oxidation.
An urgent research need is therefore to investigate
the role which such substances play for atmospheric
SO_2-oxidation(Laboratory-,field measurements,
modelling).

- Another important research need is to investigate the
  scavenging of free radicals by atmospheric droplets
  including the formation of oxidants such as H_2O_2 in
  droplets(laboratory measurements)

- Seasonal variations in rates of homogeneous and heterogeneous processes should be investigated
  (laboratory-and field measurements)

- The chemistry of NO_x and nitric acid including ammonia in the aqueous-and aerosol phase needs to be
  investigated(laboratory measurements)

- The formation of formic acid due to oxidation of
  formaldehyde should be investigated in the aqueous
  phase(laboratory measurements)

- An important research need is to investigate the role
  which formaldehyde plays in the formation of acidity
  in atmospheric cloud-and fog droplets.There is some
  recent evidence that formaldehyde reacts with sulfur
  IV to form hydroxymethanesulfonic acid(HMSA).
  As HMSA is a strong acid,SO_2 and formaldehyde may
  produce acidity in cloud-and fog water without oxidation of the sulfur IV (Richards et al.,1982).

- The contribution of natural sulfur compounds such as
  $CS_2,(CH_3)_2S$ ect.to acid rain requires an investiga-
  tion.This includes a measurement of the emissions of
  these compounds and other organic sulfur gases from
  the ocean and land surfaces(field measurements)

- The atmospheric HCl-cycle has to be investigated
  under special consideration of its interaction with
  the droplet phase

- Local and regional cycles of ammonia should be investigated in the field.Of special interest is a
  quantification of natural and anthropogenic source
  strengths of ammonia at different European locations
  and seasons.Another point is to measure the vertical,
  horizontal and temporal distribution of NH_3 and NH_4^+
  upwind,in,and downwind of large source regions of

acidic pollution(field measurements).
These field measurements should be supplemented by an
investigation of gas-liquid thermodynamic phase equilibria under special consideration of the NH3/NH$_4^+$ -
system involving deliquesced aerosols containing the
main elements of the acidity-determing martix
(laboratory-and field measurements,Brosset,1983).
Such investigations are of high priority because the
role of ammonia as a neutralizing agent must be known
if the acidity of precipitation associated with the
emissions of SO_2,NO_x,and HCl is to be understood.
The uptake of ammonia by a droplet retards its pH-
decrease causing more SO_2 and NO_x to be dissolved.
In this way the acidity of a droplet system increases
as ammonia is absorbed.

- An important research need is a parameterisation of
  the heterogeneous chemical reaction to include them
  into models.

- The generation of cloud condensation nuclei(CCN) due
  to gas-to-aerosol conversion mechanisms has to be in-
  vestigated.This is an important research need since
  the CCN play an important role for acidification of
  precipitation.On the other hand,there is strong
  evidence that over large European areas the CCN are
  to a considerable extent sulfate particles originating mainly from SO_2-oxidation and subsequent gas-to-
  particle conversion processes.

- Programmes should be designed to study the nature of
  these physico-chemical reactions in natural rain-and
  cloudwater.Of particular interest is to simulate the
  physico-chemistry of acid formation in single droplets of different size under special consideration of
  gas absorption/desorption behaviour including the
  effect of catalysts,inhibitors ect.on chemical reactions within such droplets(laboratory measurements
  and modelling).

Such investigations could help us to find answers to such key
questions as

- to what extent do local versus distant sources contribute to acid wet deposition
- what are the relative proportions of natural versus
  man made contributions to the acidity found in rainwater
- to what extent may a change of emissions in a certain
  European area influence the rain acidity in this or
  in other European areas.

7.4. Dry deposition studies

Another complex area that deserves much more attention is dry
deposition of potential acidic substances.

Of special interest is to measure in the field:

- dry deposition of NO_x and HNO_3 including ammonia.
  Special emphasis should be given to some NO_x-removal
  mechanisms suggested recently which should be operating only during nighttime(Platt et al.,1981).

- dry deposition of aerosol particles onto natural sur-
  faces.Of special interest is to investigate dry deposition over terrain with large roughness length.
  There is recent evidence that forests are important
  sinks for atmospheric aerosol particles(Höfken and
  Gravenhorst,1982).
  This includes a differentation for chemical species
  which is almost non-existent.

- the role of interception in acid deposition
  (Ulrich et al.,1979)

- There is a need for a comprehensive dry deposition
  network in Europe since dry deposition in large
  European areas seems to be at least as important as
  wet deposition for H^+-inputs to the ground.

7.5. Measurements:Analytical techniques and sampling

Some of the specified research needs imply the development of
new analytical techniques and sampling procedures.The following research needs have been specified:

- Of fundamental importance is to improve methods for
  collecting cloud water(bulk collectors and cloud
  water collectors for different droplet sizes).

- In this context it is important to develop instruments for collecting drop-free air within clouds
  sufficient for measuring gas-and aerosol concentrations.

- There is a need for improving the analytical methods
  for the determination of aerosol acidity.

- Another research need is the development of instruments capable of measuring sulfate,nitrate and
  ammonium in aerosol particles continuously.

- An important research need is to develop techniques
  for measuring ammonia,nitric acid and hydrogen peroxide in the air-and droplet phase at relatively
  low concentrations.

- A development of methods to obtain routine monitoring
  data on dry deposition of gases and particles is
  also desirable.

- There is a need to improve procedures for sampling,
  storage and analysis of dry and wet deposition samples in networks.

- There is a need for harmonisation and standardisation

of sampling and analytical procedures relevant to the
acid deposition problem.

7.6. General recommendations

As specified above,there are a series of complex interacting
factors influencing the transport,transformation and deposition of atmospheric pollutants.
In order to gain a synthesis knowledge on these aspects and
in order to obtain a useful quantitative picture,mathematical
modelling studies are required.
Modelling of acid dry and wet precipitation has the ultimate
goal to predict the changes in an observed parameter of deposition which would occur if certain changes in emissions took
place.No such models presently exist.

As the acid deposition issue is an interdisciplinary scientific field encompassing meteorological,physical,chemical and
biological aspects,a better cooperation of scientists from
different disciplines is highly desirable.

Finally,there is a need for a closer co-ordination of European
research work and a better cooperation between the various
organisations dealing with the acid deposition problem.

8. ACKNOWLEDGEMENT

The author wishes to thank the members of the CEC-Task Force
"Acid Deposition" for their valuable suggestions and useful
discussions:

      Prof.C.Brosset   (IVL,Gothenburg,Sweden)
      Ing.A.Elshout   (NV KEMA,Arnhem,The Netherlands)
      Dr.G.Gravenhorst (Lab.Glaciologie,Grenoble,France)
      Prof.A.Liberti   (Ist.Ing.Atm.,CNR,Roma,Italy)
      Dr.H.Ott   (CEC,Brussels,Belgium)
      Dr.S.A.Penkett   (AERE,Harwell,United Kingdom)
      Dr.S.Sandroni   (CEC,JRC,Ispra(Varese),Italy)
      Dr.M.Zephoris   (EERM,Magny les Hameaux,Saint Remy
                           les Chevreuse,France)

In addition,the author would like to express appreciation to
the following scientists for their critical comments and
suggestions:

      Dr.R.A.Cox   (AERE,Harwell,United Kingdom)
      Dr.H.D.Gregor   (Umweltbundesamt,Berlin,Germany)
      Dr.Ø.Hov   (NILU,Lillestrøm,Norway)
      Dr.A.S.Kallend   (CEGB,Leatherhead,United Kingdom)
      Dr.J.Paffrath   (DFVLR,Oberpfaffenhofen,Germany)
      Prof.K.Rehfuess   (Inst.Bodenkunde,Univ.München,Germany)
      Dr.R.Sartorius   (Umweltbundesamt,Berlin,Germany)
      Dr.P.Stief-Tauch (CEC,Brussels,Belgium)
      Prof.B.Ulrich   (Inst.Waldernährg.,Univ.Göttingen,Ger.)
      Dr.B.Versino   (CEC,JRC,Ispra(Varese),Italy)
      Prof.P.Warneck   (MPI,Mainz,Germany)
      RD E.Weber   (Ministry of Interior,Bonn,Germany)

Finally,the author thanks Dr.C.C.Wallén(WMO,United Nations
Environment Programme,Geneve,Switzerland) for giving per-
mission to reproduce figures 2 and 3 and Dr.L.A.Barrie
(Atmospheric Environment Service,Downsview,Ontario,Canada)
for his comments.

9. <u>REFERENCES</u>

Arndt,U.;Seufert,G.and Nobel,W.(1982)
        The contribution of ozone to the Silver-fir-desease
        A hypothesis worth to examine.Staub-Reinhaltung der
        Luft 42,Nr.6,June,pp.243-247.

Barrie,L.;Beilke,S.and Georgii,H.W.(1974)
        SO$_2$-removal by cloud-and raindrops as affected by
        ammonia and heavy metals.Proceedings:Precipitation
        Scavenging Symposium,Champaign,Illinois,USA.
        October 14-18,1974.

Barrie,L.and Georgii,H.W.(1976)
        An experimental investigation of the absorption of
        sulfur dioxide by water drops containing heavy metal
        ions.Atmospheric Environment 10,pp.743-749.

Beilke,S.and Georgii,H.W.(1968)
        Investigation on the incorporation of sulfur-dioxide
        into fog-and rain droplets.Tellus 20,pp.435-442.

Beilke,S.and Gravenhorst,G.(1978)
        Heterogeneous SO$_2$-oxidation in the droplet phase.
        Atmospheric Environment 12,No.1-3,pp.231-240.

Beilke,S.and Gravenhorst,G.(1979)
        Cycles of pollutants in the troposphere.
        Proceedings:First European Symposium on "Physico-
        Chemical Behaviour of Atmospheric Pollutants",
        Ispra(Italy),October 16-18,1979.pp.331-353.
        Editors:B.Versino and H.Ott.

Bettleheim,J.and Littler,A.(1979)
        Historic Trends in Sulfur Oxide Emissions in Europe
        since 1865.CEGB Report PL-GS/E/1/79.
        (cited in:KEMA-report on "Effects of SO$_2$ and its
        Derivatives on Health and Ecology",volume 2,
        November 1981).

Böttger,A.;Ehhalt,D. and Gravenhorst,G.(1978)
        Atmosphärische Kreisläufe von Stickoxiden und
        Ammoniak.Bericht der Kernforschungsanlage Jülich,
        No. 1558,November 1978.

Boulaud,D.;Bricard,J. and Madelaine,G.(1978)
        Aerosol growth kinetics during SO$_2$-oxidation.
        Atmospheric Environment 12,No.1-3,pp.171-178.

Brosset,C.(1979)
    Possible changes in aerosol composition due to
    departure from equilibrium conditions during sampling.
    Paper presented during ACS/CSJ Chemical Congress,
    Honolulu,Hawaii,April 1-6,1979.

Brosset,C.(1983)
    Characterization of acidity in natural waters.
    Paper presented at Workshop on "Acid Deposition",
    Berlin,Reichstag, 9 September 1982. This volume.

Calvert,J.G.;Su,F.;Bottenheim,J.W.and Strausz,O.P.(1978)
    Mechanisms of the homogeneous oxidation of sulfur
    dioxide in the troposphere.Atmospheric Environment
    12,No.1-3,p.197.

Carbonelle,J.(1981)
    Preliminary results on contribution of Mt.Etna and
    Mt.Stromboli to atmospheric SO_2 and CO_2.Paper pre-
    sented at 2nd meeting of Working Party 4(Pollutant
    Cycles),Aussois(France),April 28-29,1981.
    Project COST 61a bis.

Chamberlain,A.C.(1975)
    The movement of particles in plant communities.
    In:Vegetation and the atmosphere.Edited by Monteith,
    Academic press,pp.155-201.

Charlson,R.J.and Rodhe,H.(1982)
    Factors controlling the acidity of natural rain-
    water.Nature,Vol.295,No.5851,pp.683-685.

Cox,R.A.(1976)
    Review paper:Homogeneous gas phase oxidation of
    sulfur dioxide and other atmospheric sulfur gases.
    Paper presented at Technical Symposium,Ispra(Italy),
    November 1976.Project COST 61a.

Cox,R.A.(1982)
    Discussion during meetings of pilots of project
    COST 61a bis,June 1st,Brussels.

Cox,R.A.and Penkett,S.A.(1983)
    Formation of atmospheric acidity.Paper presented at
    Workshop on "Acid Deposition",Berlin,Reichstag,
    9 September 1982. This volume.

Cullis and Hirschler(1980)
    (cited in: Herron,M.M.(1982)
    Impurity Sources of F^-,Cl^-,NO_3^- and $SO_4^=$ in Greenland
    and Antarctic Precipitation.Journal Geophysical
    Research,Vol.87,No.C 4,April 20,1982.p.3053.

Daum,P.H.;Schwartz,S.E.and Newman,L.(1982)
    Studies of the gas and aqueous phase composition of
    stratiform clouds.Paper presented at 4th Internat.

Conference on Precipitation Scavenging,Dry Deposition and Resuspension,Santa Monica,California,USA,
29 November - 3 December,1982.

Delmas,R.;Baudet,J.and Servant,J.(1980)
Emissions and concentrations of H_2S in the air of
tropical forest of the Ivory Coast and of temperate
regions(France).Paper presented by A.Marenco at 1st
meeting of Working Party 4(Pollutant Cycles),Berlin,
May 6-7,1980.Project COST 61a bis.

Delmas,R.and Gravenhorst,G.(1983)
Background precipitation acidity.Paper presented at
Workshop on "Acid Deposition".Berlin,Reichstag,
9 September 1982.This volume.

Eggleton,A.E.J.and Cox,R.A.(1978)
Homogeneous oxidation of sulphur compounds in the
atmosphere.Atmospheric Environment 12,Vol.1-3,
pp.227-230.

Elshout,A.(1982)
Discussion during 2nd meeting of the Task Force on
Acid Deposition,June 1st,1982,Brussels.

Elshout,A.and Römer,F.G.(1983)
Measurements of cloud water composition.Paper pre-
sented during Workshop on acid deposition,Berlin,
Reichstag,September 1982.This volume.

Fjeld,B.(1976)
Forbruk av fossilt brensel i Europa og utslipp av
SO_2 i perioden 1900-1972.Norwegian Institut for Air
Research,Teknisk Notat No. 1/76.

Fowler,D.(1980)
Removal of sulphur and nitrogen compounds from the
atmosphere in rain and by dry deposition.Proceedings
International Conference on Ecological Impact of
Acid Precipitation.Sandefjord,Norway,March 11-14,
1980.

Garland,J.A.(1982)
Field measurements of the dry deposition of small
particles to grass.Proceedings Conference on Depo-
sition of Atmospheric Pollutants.Oberursel(Germany),
November 9-11,1981.Reidel Publication,pp.9-16.
Editors:Georgii and Pankrath.

Georgii,H.W.and Perseke,C.(1979)
Some results on wet and dry deposition of sulfur
compounds.Proceedings 1st European Symposium on
Physico-Chemical Behaviour of Atmospheric Pollutants,
Ispra(Italy),October 16-18,1979.
Editors:B.Versino and H.Ott

Granat,L.(1972)
   On the relation between pH and the chemical composition in atmospheric pollution.Tellus 24,pp.550-560,
   1972.

Gravenhorst,G.(1978)
   Maritime sulfate over the North Atlantic.
   Atmospheric Environment 12,No.1-3,pp.707-713.

Gravenhorst,G.;Beilke,S.;Betz,M.and Georgii,H.W.(1980a)
   Sulfur dioxide absorbed in rain water.Nato Conference
   Series 1,Vol.4:The effect of acid precipitation on
   terrestrial ecosystems.Editors:Hutchinson,T.C. and
   Havas,M.(Plenum Publ.Corp.,1980),pp.41-55.

Gravenhorst,G.;Janssen-Schmidt,T.;Ehhalt,D.and Röth,E.
   The influence of clouds and precipitation on the
   vertical distribution of SO2 in a one-dimensional
   steady-state model.Atmospheric Environment 12,
   No.1-3,pp.691-693(1978)

Gravenhorst,G.;Müller,K.P. and Franken,H.(1979)
   Inorganic nitrogen in marine aerosols.Gesellschaft
   für Aerosolforschung 7,pp.182-187.

Gravenhorst,G.;Perseke,C.;Rohbock,E.and Georgii,H.W.
   Untersuchungen über die trockene und feuchte Deposition von Luftverunreinigungen in der Bundesrepublik
   Deutschland.Forschungsbericht:Institut für Meteorologie der Universität Frankfurt, 1980b

Höfken,K.D.;Georgii,H.W. and Gravenhorst,G.(1981)
   Untersuchungen über die Deposition atmosphärischer
   Spurenstoffe an Buchen-und Fichtenwald.Berichte des
   Institutes für Meteorologie der Universität Frank-
   furt,Nr.46.

Höfken,K.D.and Gravenhorst,G.(1982)
   Deposition of atmospheric aerosol particles to beech-
   and spruce forest.
   In:Deposition of Atmospheric Pollutants,Reidel Publ.
   Comp.,pp.191-194.
   Editors:Georgii and Pankrath.

Hutchinson,T.C.and Havas,M.(1980)
   The effects of acid precipitation on terrestrial
   ecosystems.Plenum Publ.Corp.,Nato Conf.Series 1,
   Vol.4.

International Conference on the Effects of Acid Precipi-
   tation.Telemark,Norway,June 14-19,1976.
   Proceedings:Ambio No.5-6(1976)

ISSA,Workshop conclusions on dry deposition.
   In:Atmospheric Environment 12,No.1-3,p.14(1978)

Jaeschke,W.(1980)
    Sulfur emissions from Mt.Etna.Paper presented at 1st
    meeting of Working Party 4(Pollutant Cycles),6-7May,
    1980.Berlin.Project CUST 61a bis.

Janssen-Schmidt,T.;Röth,E.P.;Varhelyi,G.and Gravenhorst,G.
    Anthropogene Anteile am atmosphärischen Schwefel-und
    Stickstoffkreislauf und mögliche globale Auswirkungen
    auf chemische Umsetzungen in der Atmosphäre.
    Berichte der Kernforschungsanlage Jülich,Nr.1722.

Judeikis,H.S.and Wren,A.G.(1978)
    Laboratory measurements of NO-and NO_2-deposition onto
    soil and cement surfaces.Atmospheric Environment 12,
    p.2315.

Junge,C.and Scheich,G.(1971)
    Determination of the acid content of aerosol par-
    ticles.Atmospheric Environment 5,pp.165-175.

Kallend,A.S.;Marsh,A.R.;Pickles,J.H.and Proctor,M.V.(1982)
    Acidity of rain in Europe.Paper accepted for publi-
    cation in Atmospheric Environment.

Kallend,A.S.(1983)
    Trends in the acidity of rain in Europe: a re-exami-
    nation of European atmospheric chemistry network
    data.Paper presented at Workshop on Acid Deposition,
    Berlin,Reichstag, 9 September 1982. This volume.

Kallend,A.S.;Marsh,A.R.;Glover,G.M.;Webb,A.H.;Moore,D.;
    Clark,P.A.;Fisher,B.E.A.;Dear,D.J.A.;Lightman,P.
    and Laird,C.K.(1981)
    Studies of the fate of atmospheric emissions in
    power plant plumes over the North Sea.
    Proceedings: 2nd European Symposium on Physico-
    Chemical Behaviour of Atmospheric Pollutants,Varese
    (Italy), 29 September - 1 October,1981.
    Editors:B.Versino and H.Ott.

KEMA Report:Effects of SO_2 and its Derivatives on Health
    and Ecology.Summary Work.Group Rep. Vol.1,Nov.1981.

Koch,C.(1979)
    Experimentelle Untersuchungen der trockenen und
    nassen Ablagerung von atmosphärischen Schwefelver-
    bindungen SO_2 und Sulfat(Experimental investigation
    of dry and wet deposition of atmospheric compounds
    SO_2 and sulfate),Diplomarbeit,Institut für Meteoro-
    logie,Universität Frankfurt.

Last,F.T.;Likens,G.E.;Ulrich,B.and Walloe,L.(1980)
    Acid precipitation-progress and problems.
    Proceedings:International Conference on Ecological
    Impact of Acid Precipitation,Sandefjord,Norway,
    March 11-14,1980.

Liberti,A.(1982)
    Discussion during 2nd meeting of Task Force "Acid
    Deposition",Brussels,June 1,1982.

Meurrens,A.and Lenelle,Y.(1976)
    Sulfites et sulfates dans la pluie-role de la pluie
    dans l'elimination du SO_2 atmospherique.Paper dis-
    tributed at Ispra,November 1976.COST 61a.

Miller,J.M.and Yoshinaga,A.M.(1981)
    The pH of Hawaiian precipitation.Geophysical Research
    Letters 7.

Müller,J.;Reuver,H.and Jost,D.(1981)
    Measurements of F^-,Cl^-,NO_3^-,and $SO_4^=$-ions in rainwater
    and particulate matter by aid of ionic-chromatogra-
    phy.Proceedings: 2nd European Symposium on Physico-
    Chemical Behaviour of Atmospheric Pollutants,Varese
    (Italy),29 September - 1 October,1981.pp.440-448.
    Project COST 61a bis.

Nguyen,B.C.;Bonsang,B.;Gaudry,A.and Lambert,G.(1980)
    Gaseous marine sulphur compounds in the atmospheric
    sulphur cycle.Paper presented at 1st meeting of
    Working Party 4(Pollutant Cycles),Berlin,May 6-7,
    1980.Project COST 61a bis.

Penkett,S.A.;Jones,B.M.R.;Brice,K.A.and Eggleton,A.(1979)
    The importance of atmospheric ozone and hydrogen
    peroxide in oxidizing SO_2 in cloud-and rain water.
    Atmospheric Environment 13,pp.123-138.

Platt,U.;Perner,D.;Schröder,J.;Kessler,C.and Toenissen
    (1981)
    The diurnal variation of NO_3.Paper presented at 2nd
    meeting of Working Party 4(Pollutant Cycles),Aussois
    (France),April 28-29,1981.Project COST 61a bis.

Prinz(1982)
    Personal communication to S.Beilke

Rehfuess,K.E.(1981)
    Über die Wirkungen der sauren Niederschläge in
    Waldökosystemen.Forstwissenschaftliches Zentral-
    blatt,100.Jahrgang,H.6,S.363-381.

Richards,L.W.;Anderson,J.A.;Blumenthal,D.L.;Mc Donald,J.A.
    Kok,G.L. and Lazrus,A.L.(1982)
    Hydrogen peroxide and sulfur(IV) in Los Angeles
    Cloud Water.Paper presented at SCADDER Conference,
    29 November - 3 December 1982,Santa Monica,USA.

Rodhe,H.;Crutzen,P.and Vanderpol,A.(1981)
    Formation of Sulfuric and Nitric Acid in the Atmos-
    phere during Long-range Transport,Tellus 33,p.132

Römer,F.G.;Viljeer,J.W.;Van den Beld,L.;Slangewal,H.J.;
    Veldkamp,A.A. and Reijnders,H.F.R.(1983)
    Preliminary measurements from an aircraft into the
    chemical composition of clouds.Paper presented at
    Workshop on "Acid Deposition",Berlin,Reichstag,
    9 September 1982.This volume.

Sandroni,S.(1982)
    Discussion during 2nd meeting of Task Force "Acid
    Deposition",Brussels,June 1st,1982.

Schurath,U.(1980)
    Luftchemisches Verhalten von NO_x.Ergebnisse der VDI-
    Arbeitsgruppe "Luftchemie" in der VDI-Kommission
    Reinhaltung der Luft,S.36-42.

Sequeira,R.(1982)
    An Assessment Based on Acid-Base Considerations.
    JAPCA,March 1982,Vol.32,No.3,pp.241-245.

Ulrich,B.;Mayer,R. and Khanna,C.K.(1979)
    Deposition von Luftverunreinigungen und ihre Auswir-
    kungen in Waldökosystemen im Solling.
    Schriften aus der forstlichen Fakultät der Universi-
    tät Göttingen und der niedersächsichen Forstlichen
    Versuchsanstalt,Band 58,Frankfurt.

Ulrich,B.(1983)
    Effects of acid deposition.Paper presented at Work-
    shop on Acid Depositinn,Berlin,Reichstag, 9 September
    1982.This volume.

Umweltbundesamt Report: Luftreinhaltung 81 -Entwicklung,
    Stand,Tendenzen.Erich Schmidt Verlag,Berlin,1981.

Varhelyi,G.and Gravenhorst,G.(1981)
    An attempt to estimate biogenic sulfur emissions into
    the atmosphere.Idöjaras 85,3,p.126.

Weiss,R.F.(1981)
    The temporal and spatial distribution of tropospheric
    nitrous oxide.Journal Geophysical Research 86,C8,
    pp.7185-7195.

Winkler,P.(1977)
    Automatic analyser for pH and electrical conductivity
    of precipitation.In:WMO Bulletin,Nr.480,Geneve,
    pp.191-196.

Winkler,P.(1983)
    Trend developments of precipitation pH in central
    Europe.Paper presented at Workshop on "Acid Deposi-
    tion",Berlin,Reichstag, 9 September,1982.
    This volume.

WORKSHOP: Advisory Workshop to Identify Research Needs
     on the Formation of Acid Precipitation.Alta,Utah,USA,
     August 22-25,1978.
     Proceedings in:EA-1074,Special Study Project WS-78-98
     Prepared by SIGMA Research Inc.
     2950 George Washington Way,Richland,Washington 99352
     Editor:D.H.Pack
     Prepared for:Electric Power Research Institute(EPRI),
     Project Manager:C.Hakkarinen.

WORKSHOP: Final Report on the Source-Receptor Relation-
     ship in Acid Precipitation,La Jolla,California,USA.
     January 1982.Report Physical Dynamics Inc.,
     PD-LJ-82-268R.

Wallen,C.C.(1961)
     A preliminary evaluation of the WMO/UNEP precipita-
     tion chemistry data.MARC report No.22.

EFFECTS OF ACID DEPOSITION

B. ULRICH
Institut für Bodenkunde und Waldernährung der Universität
D-3400 Göttingen, Büsgenweg 2

Summary

From the view point of the receiving ecosystem, a distinction is made
between precipitation deposition and interception deposition. Long
term mean values for both kinds of deposition are given for a
deciduous and a coniferous forest in Central Europe (Solling region).
Interception deposition exceeds precipitation deposition especially
in the coniferous forest. The acidity deposited is to more than 75 %
due to SO_2, NO_x contributes less than 20 %. Only less than 22 % of the
acidity is neutralized before reaching the ecosystem. Less than 18 %
of the acidity is buffered in the ecosystem down to 1 m soil depth.
The main fraction of the acidity passes through the ecosystem and will
finally acidify the hydrosphere. The two aspects of soil acidification
are discussed shortly: leaching of nutrients and release of toxins.
Heavy metals from deposition accumulate in the biotic part of the
ecosystem. From a balance consideration, taking into account emission
rates, deposition rates and the kind and rates of possible reactions
of strong acids in the ecosphere, it is concluded that the emission of
strong acid formers like SO_2 and NO_x leads to the poisening of the
ecosphere.

Grouping of air pollutants according to effects

Most of the elements supplied from the air to forest ecosystems have a
natural source which may be outweighted by anthropogenic sources. The
following grouping is relevant in Central Europe which is under strong
influence of many anthropogenic sources with contributions from soil dust
and sea salt:
- neutral salts: e.g. Na^+, Cl^- (from sea salt)
- nutrients: Mg^{2+}, Ca^{2+}, NH_4^+, NO_3^-, SO_4^{2-}, micronutrients
- acid formers: SO_2, NO_x, Cl_2
- potential toxins: SO_2, HF, heavy metals, As, Se, hydrocarbons
- oxidant formers: NO_x, gaseous organic compounds.

This list is by far not complete. It is evident, however, that a full understanding of effects requires first of all a complete knowledge of the load.
This knowledge is nowhere available.

Deposition processes

The following deposition processes can be distinguished:
- wet deposition by rain or snow (A)
- sedimentation (gravimetric fall) of particles (B)
- impaction of aerosols (C)
- impaction of mist, fog, cloud droplets (C)

- absorption of gases (example: SO_2) (D)
  - on wet surfaces like foliage, bark, wet snow
  - inside stomata
    - on cell walls
    - on to mesophyll and palisade cell surfaces
- reemission

Wet deposition of rain or snow is not influenced quantitatively by the receiving surface. Deposition of particles (sedimentation, impaction) is strongly dependent on particle size. Fog, mist, cloud droplet could be regarded as large particles and are very efficiently captured by a canopy making the rate of deposition dependent mostly on the wind speed and thus on the exposure of the receiving surface to wind action. Deposition of gases varies with kind and state of the receiving surface.

The deposition can be looked at from two different points of view: either from the depositing compound or from the receiving surface. From the point of view of depositing compounds wet and dry deposition are distinguished. From the point of view of the receiving surface it makes sense to distinguish between precipitation deposition and interception deposition. These terms are shortly defined in the following comparison:

	Process	view point of depositing compound	view point of receiving surface
A	precipitation of rain or snow "particles" with dissolved (soluble) or undissolved (unsoluble) content	wet deposition	precipitation
B	precipitation of particles other than rain or snow according to gravity	dry deposition	deposition
C	impaction of aerosols including fog and cloud droplets according to air or Brownian movement [+)]		interception
D	dissolution of gases on wet surfaces (with subsequent chemical reactions)		deposition (ULRICH et al. 1979)

[+)] the impaction of fog and cloud droplets is sometimes grouped under wet deposition.

It is evident that the processes summarized under the term interception deposition depend upon size, kind and (chemical) state of the receiving surface. This has important consequences regarding the measuring techniques.

After deposition chemical reactions can occur. The following reaction types can be distinguished:
- acid production (e.g. SO_2)
- proton buffering (e.g. cation exchange on cell walls)
- precipitation or dissolution, depending on pH (e.g. heavy metals)
- reactions with living plant cells (reactions with cell membranes, uptake into cells, assimilation in the cell metabolism)

- formation of gaseous compounds (e.g. sulfides)

After deposition a translocation could also occur within the canopy or by leaching by rain from the outside as well as from the inside of leaves and needles. Much of the deposited material will by this process be transferred from the canopy to the ground.

Forest canopies as measuring devices for interception deposition

No artificial collector surfaces can be expected to give realistic values of interception deposition. The surface of the unchanged receiving system seems to be the only acceptable sampling device, which can be assumed to give realistic interception deposition values. This includes even the unchanged metabolism of plant canopies, since the dry deposition of gases depends upon the chemical reactions following absorption (installing infinite sinks).

The importance of dry deposition (for instance as compared with wet deposition) varies very significantly between areas. Deposition on wet canopy may be very rapid in the simultaneous presence of SO_2 and ammonia while it may be unimportant in a remote area where the throughfall is quite acid. Areas where fog or low clouds are frequent may receive large additional input by the interception of droplets.

Measurements based on throughfall measurements use the living forest canopy with its variation in surface and metabolism as receiving surface. In this paper the approach of MAYER and ULRICH (1974) taking a forest canopy as measuring device for determining interception deposition is used. The approach is based on the flux balance of the forest canopy (ULRICH et al. 1979) or of the whole ecosystem (MAYER, 1981), depending upon the mobility of the air pollutant considered. This approach is in principle applicable to any air pollutant, including organic micropollutants. In the flux balance equation used, the function of the canopy or the whole vegetation cover to act as sink or source for the air pollutant considered has to be expressed quantitatively as annual rate.

A forest canopy can be looked at as a compartment which is connected with its environment by inputs and outputs.

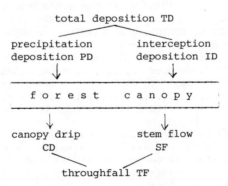

The input, total deposition TD, consists of precipitation deposition PD and interception deposition ID, where PD can be measured with simple precipitation samplers and chemical analysis. The output occurs as canopy drip and stem flow, which yield together the throughfall TF. Also TF can be measured by collecting the precipitation beneath the canopy. The forest canopy may itself exert a sink or source function Q. If X denotes a distinct chemical compound or ion and if annual rates are considered, the following balance equation applies:

$$TF_X = PD_X + ID_X + Q_X \tag{1}$$

If PD_X and TF_X are measured, knowledge of Q_X is necessary in order to solve equation 1 for the interception deposition of X, ID_X.
The canopy can act as a source in the following cases:
- leaching of ions from senescent leaves mainly in autumn
  $X \in \{Na, Mg, Ca, Cl, SO_4\}$
- cation exchange in the leaf tissue (see also below)
  $X \in \{Ca, Mg\}$
- leaching of ions throughout the growing season due to metabolic processes
  $X \in \{K, Mn\}$
- dissolution of undissolved matter in deposited particles
  $X \in \{Al, heavy metals\}$
The canopy can act as a sink in the following cases:
- assimilation, i.e. uptake into cell metabolism
  $X = NH_4^+$. At high rates of sulfur deposition the fraction of S assimilated can be neglected
- cation exchange in the leaf tissue (see also above, exchange of H^+ against Ca and Mg)
  $X = H^+$
- storage of particles
  $X \in \{Al, heavy metals\}$
- precipitation of dissolved ions
  $X \in \{Al, heavy metals\}$

Input, storage and output of acids and bases in a case study

The approach has been applied to two forest ecosystems, a 120 yrs. old beech (Fagus silvatica) stand and a 85 yrs. old Norway spruce (Picea abies) stand. Both stands lie close together in the Solling mountains, $9^\circ 30'$ east of Greenwich and $51^\circ 40'$ north, on a plateau 500 m above sea level. The site is described in detail by ELLENBERG (1971). The data used are presented in detail by ULRICH et al. (1979) and MATZNER et al. (1982).
Following Brønsted's definition of acids, the acidity to be considered consists of hydrogen ions (protons, H^+) and cation acids like NH_4^+, Fe^{n+}, Mn^{2+} and Al^{3+} ions. NH_4^+ produces a proton when it is taken up by cells and used for amino acid formation. The other cation acids, including heavy metals, can produce protons if they react with water and form hydroxo compounds. In the existing acidified environment carbonic acid is expelled and plays no role. As basic component nitrate ions have to be taken into account. If nitrate is taken up by cells and transferred into amino acids, a proton is consumed. The accumulation of Ca and Mg ions in exchangeable or complexed form in the plant is connected with an equivalent production of protons by dissociation of acidic groups. It represents therefore the accumulation of basicity in the plant which is balanced by an equivalent accumulation of acidity in the soil.
In table 1 the input/output balance for the acids and bases is given. The data are mean values for the measuring period from 1969 to 1979, in some cases the measuring period started later (1973).
As can be seen from table 1, interception deposition amounts to a considerable fraction of total deposition. Both stands lie close together,

Table 1: Input/output balance of a beech and a spruce forest ecosystem in Central Europe in respect to acidity

	H	NH_4	cation acids			sum acids	S	NO_3	Cl	sum acid formers
			$\frac{1}{3}$ Fe	$\frac{1}{2}$ Mn	$\frac{1}{3}$ Al					
	------------------------------------- kiloequivalents per ha and per year -------------------------------------									
B E E C H										
Precipitation deposition	0.82	0.85	0.05	0.01	0.13	1.86	1.48	0.57	0.48	
Interception deposition	0.88	0	0.05	0.01	0.14	1.08	1.58	0.14	0.39	
Total deposition	1.70	0.85	0.1	0.02	0.27	2.94	2.85[1]	0.71	0.22[1]	3.78
Seepage output	0.48	0.02	0.01	0.19	1.70	2.40	2.26	0.12		
Input minus output						+0.54	+0.80[2]	+0.54		
S P R U C E										
Interception deposition	2.87	0.23	0.07	0.07	0.21	3.45	3.80	0.50	0.54	
Total deposition	3.69	1.08	0.12	0.08	0.34	5.31	5.03[1]	1.07	0.22[1]	6.32
Seepage output	0.38	0.01	0.01	0.33	4.31	5.04	4.47	0.85		
Input minus output						+0.27	+0.81[2]	+0.22		

1) Corrected for sea salt, see text
2) Not corrected for sea salt

the climatical conditions and the exposure to wind are therefore almost the same. The difference in interception deposition between beech and spruce reflects thus only the difference of the canopies to interact with air constituents. The high interception deposition of spruce seems to be caused by dry deposition of SO_2 on wet canopy surfaces during autumn and winter, especially during fog situations, and by the capture of cloud droplets.

As pointed out in detail elsewhere (ULRICH 1983a), the canopy buffers protons by transferring Ca and Mg bicarbonates into sulfates under formation of CO_2 and H_2O; the reaction is mediated by cation exchange in the cell wall space. The annual rate of this reaction has been calculated for the stands investigated, the data are given in ULRICH (1983b, fig.1). In the mean of the time period considered, this buffering amounts to 0.42 keq ha^{-1} yr^{-1} for beech and to 0.68 keq for spruce.

If sulfate accumulates in the soil in the form of $AlOHSO_4$, this corresponds to the accumulation of an undissolved acid. During dissolution $AlOHSO_4$ can form H_2SO_4 and $Al(OH)_3$. Therefore the S-balance is given separately in table 1.

Part of the sulfate deposition is due to sea salt, this part is not creating acidity. The same is true to a much greater extent for chloride. In table 1, the values given for total deposition of S and Cl have been corrected for the sea salt fraction assuming that the total deposition of Na (as given in MATZNER et al. 1982) is due to sea spray. The values of total deposition allow than a comparison between the sum of acids deposited and the sum of acid formers SO_2, NO_x and Cl, deposited as sulfate, nitrate and chloride. 78 % (beech) to 84 % (spruce) of the acidity produced by the acid forming gases is deposited in the ecosystems, that means that only 16 to 22 % are neutralized before reaching the ecosystem. The acidity deposited is mainly (75 to 80 %) caused by SO_2, NO_x contributes 17 to 19 %, Cl 3 to 6 %.

In respect to the kind of the acid, protons dominate in deposition whereas aluminum ions dominate in the seepage output from the soil. The seepage output of NH_4 ions is almost zero. Since the main fraction of NH_4 remaining in the ecosystem is transferred to organic bound nitrogen, this means that the cation acid NH_4 reacts completely forming an equivalent amount of protons.

According to the input/output balance (last row), from the acidity received only 18 % (beech) and 5 % (spruce) are buffered in the ecosystem down to 1 m soil depth. The main fraction of the acidity deposited passes through the ecosystem and will acidify seepage conductors or ground and spring waters. This situation must be expected for all ecosystems where the soil has lost the exchangeable basic cations to a percentage of less than 5 to 10 % of the cation exchange capacity. This applies to many forest soils in Central Europe. It cannot be questioned that the input of acidity into the ecosphere will finally lead to the large scale acidification of ground waters. This acidification implies the presence of Al and heavy metal ions in the water. Acidification damage to lakes and watercourses has been reported from most parts of the world. Those that are hardest hit are Sweden, Norway, the U.S. and Canada. They are also reported from Central Europe, but due to the lack of lakes in main forest regions and the mixing of forest and agricultural land use this problem has not yet achieved so much attention.

As discussed already nitrate represents the only base being deposited if the pH in the deposition is below 5. The input/output balance of nitrate shows that the beech ecosystem is much more effective in utilizing nitrate for amino acid production. Beech can grasp 76 % of the nitrate which has been deposited, spruce only 2o %. The character of nitrate as a base remains if it is denitrified to N_2, NO or N_2O. The type of reaction of nitrate inside the ecosystem has therefore no effect on the acid/base balance.

Soil acidification

Soil acidity has a quantitative (capacity) and a qualitative (intensity) aspect. The capacity can be defined as the equivalent sum of acids which can be neutralized by addition of a strong base to pH 7 or 8. The intensity is expressed as the thermodynamic activity of the protons and is measured as pH value, it is determined by the strength of the acids controlling the proton activity in the soil solution. Soil acidification means either a loss of basicity or an increase of acidity, both considered as capacity terms. Only in special buffer ranges a loss of basicity or an increase of acidity is connected with changes in pH. This is especially true when the exchangeable bound cations (mainly Ca^{2+}, to some extent Mg^{2+}) are replaced by Al ions and leached. It must clearly stated that no change in pH does not mean that soils did'nt acidify.

The acid load from deposition in forest soils in Central Europe may vary between 3.5 and 7 keq ha^{-1} yr^{-1}. The corresponding annual loss of exchangeable Ca can approach the cation exchange capacity (CEC) due to 0.5 to 1 % clay in a soil layer of 10 cm thickness. That means that within 50 years, the exchangeable Ca may be completely leached from a soil containing 5 to 10 % clay to a depth of 50 cm. Since the acid deposition increased rapidly with the beginning industrialization, it is obvious that a considerable soil acidification must have occured due to acid deposition during the last century and especially since 1950 in Central Europe. The ecosystems presented in table 1 have already passed through the phase of leaching of exchangeable cations. After this phase the accumulation of acids becomes very limited in size, the acidity is therefore just passing through the ecosystem as demonstrated in table 1.

Soil acidification has two aspects: the leaching of nutrients and the release of toxins. The leaching of nutrients results in nutrient deficiency. The toxins released are either cation acids like Mn, Al, Fe and heavy metal ions, or water soluble phenols. The cation acids are mainly released in the mineral soil, the phenoles in the organic top layer. The presence of toxic substances means that tolerance mechanisms against the toxins play a deciding role in competition. This is true for plants as well as for decomposer organisms. Soil acidification leads therefore to drastic changes in the species composition of the decomposers and the vegetation. Under natural conditions, soil acidification goes so slowly that we find ecosystems adjusted to different degrees of soil acidity neighbouring each other as ecosystems being in a steady state. Even under the load of acid deposition, the ecosystem succession caused by soil acidification should be expected to take decades up to more than a century. The development becomes easily visible and reaches the features of a catastrophe when the dominating tree species disappear from the ecosystem. In other continents with a continuous plant cover since the tertiary age, species evolution and ecosystem evolution had time enough to develop tree species and forest ecosystems which are adapted to strongly acidified soils, their resilience is very low however. It seems that the tree species present in Central Europe are not adjusted to highly acidified soils. Such soils will therefore carry a treeless vegetation of low productivity.

The rapid tree and forest die-back happening now in Central Europe cannot be explained solely by soil acidification, however. The die-back proceeds too fast and lays hold of trees and forests on not acidified soils also. It is self-evident that all aspects of air pollution should contribute to the forest damage. The other aspects of air pollution include mainly direct damage by SO_2 and ozone and toxic effects of heavy metals. In the context of this paper direct effects of SO_2 and ozone as well as of organic micropollutants will not be treated.

Mobilization and accumulation of heavy metals

In table 2 data are presented from the spruce ecosystem in the Solling about the deposition and storage places of Al and the heavy metals investigated.

According to these data half to 2/3 of the deposition of Cr, Co, Ni and Cu is first accumulating in the crown. Part of these heavy metals passes over to the soil with litter fall, but 20 to 40 % of the deposition of these metals remain in the bark of branches and stem. Under continuous deposition, a hundred year old stand should have accumulated in the bark 6 kg Cr, Ni, and Pb each and 24 kg Cu. It is obvious that such an accumulation of toxic metals, in combination with low pH in water films covering leaf and bark surfaces, must be expected to lead to injury of living cells and thus to damages of leaves, buds, bark, and branches. Bark necrosis is indeed visible on many trees, especially on upper crowns and single trees, which are exposed to higher rates of deposition. Another symptom which is probably caused by heavy metal toxicity is the continuous loss of green leaves and short twigs throughout the vegetation period on beech, oak and spruce. This phenomenon was registered during the last four years in many forests, together with the die-back of trees in cities. It has to be assumed on the base of these data that heavy metal toxicity plays an important role in tree damages and die-back inside and outside of forests.

Table 2 shows further that, with the exception of Zn and Cd, the heavy metals accumulate in the organic top layer of the soil. Here the decomposition of the leaf and part of the root litter takes place; this horizon is also rooted. It contains already 25 kg Pb/ha, under continuous deposition this will increase to more than 50 kg Pb/ha. It cannot be expected that with such amounts present these metals would'nt find their way into cells and cause injury to decomposers and roots.

According to table 2, the highly acidified mineral soil accumulates only Pb, whereas Ni, Cu and Cd are leached with the seepage water and thus transferred to the hydrological cycle. The present situation is that the heavy metals are accumulating in the biotic part of the ecosystem, whereas they are leached from the mineral part. Such a situation must be characterized as poisoning of the biosphere.

The ecological meaning of the acidification of the environment

SO_2 and NO_x form, in the presence of liquid water and oxygen, strong mineral acids (H_2SO_4 and HNO_3). These strong acids have the tendency to form weaker acids by reacting with bases. In the air, originating from air pollution, and in the ecosphere three different groups of bases are available:
1. Alcali- and earth alcali-carbonates and -silicates: The reaction with strong acids leads to the formation of alcali- and earth alcali-sulfates and -nitrates and the weak acids carbonic acid and silicic acid.

Table 2: Deposition and storage places of aluminium and heavy metals in the spruce ecosystem in the Solling (according to MAYER 1981, 1983)

	Al	Cr	Fe	Co	Ni	Cu	Zn	Cd	Pb
				kg element per ha and year					
total deposition	2.8	0.17	2.1	0.02	0.14	0.66	1.7	0.02	0.73
from dep. accumulated in stand	0	0.13	0.01	0.008	0.1	0.43	0.18	0	0.27
from dep. accumulated in bark and wood	0	0.06	0.001	0.004	0.06	0.24	0.04	0	0.06
storage change in soil top organic	-2.4	+0.08	+9	+0.006	+0.07	+0.31	-0.88	+0.002	+0.55
storage change in mineral soil	-21	-0.001	-8		-0.08	-0.06	+0.002	-0.01	+0.1
output with seepage water	24	0.006	0.16		0.07	0.11	2.4	0.03	0.013

At pH-values below 7 to 5 these acids are transferred to water and CO_2 and SiO_2, respectively. The protons of the strong acids are in this case transferred to the weakest acid existing: water. This corresponds to a neutralization. The acids disappear, the salts formed contain nutrients which are toxic only at very high concentrations.

The presence of carbonates in forest soils is very limited, this buffer substance plays a role almost only if it is added to the soil by fertilization. Silicates are, on the contrary, very widespread in soils. However, their reaction with strong acids is kinetically limited. The reaction rate will usually not exceed 1 kmol H^+ ha^{-1} yr^{-1}, it may in soils rich in easily weatherable silicates reach 2 kmol. In managed forests, this natural rate of acid buffering has been more or less used up for neutralizing the acidity caused by the export of wood from the forest ecosystem (ULRICH 1983a). In general one can say that any additional acid input cannot be buffered by silicate weathering.

2. NH_3: The rate of emission of ammonia is estimated to 2 to 14 kg N ha^{-1} yr^{-1} for Europe; in the Solling the mean annual deposition was 12 kg NH_4-N/ha. NH_3 is an important reactant in the atmosphere; with H_2SO_4 and HNO_3 ammonium sulfates and -nitrates are formed. As can be seen from table 1, however, NH_4 is used completely in the ecosphere for the formation of proteins and the proton is released again. The buffering by the formation of ammonium sulfates and nitrates in air is therefore only an intermediate process and should not be taken into account in an overall balance.

3. Oxides and silicates of aluminum and heavy metals: Heavy metals enter the air as pollutants from industrial processes and coal burning (fly ash as well as in gaseous form). In soil minerals, aluminum, iron and manganese are dominating. If these compounds react with strong acids, the ionic forms of the metals are released as cation acids. These cation acids are more or less toxic at already low concentrations; some of them are nutrients (trace elements, micronutrients).

In addition the strong mineral acids can react with organic substances. For an overall balance of the ecosphere this again need not to be taken into account, since the organic substances are finally oxidized to CO_2 and H_2O. It may become of great importance, however, if the intermediate formed organic substances are toxic. Such toxic substances are oxidants, which are formed in the air, and water soluble phenols, which are formed in the soil.

The acidity caused by NO_x can be neutralized in the ecosphere by the formation of protein and by denitrification of nitrate. The role of NO_x in the acidification of the ecosphere is therefore more complex as the role of SO_2. NO_x plays, however, a key role in the formation of oxidants. Oxidants seem to play an increasing role in causing acute leaf injury.

From this consideration it follows that the emission of strong acid formers like SO_2 and NO_x leads to the poisening of the ecosphere. Together with the emission of heavy metals and organic micropollutants already the air and the rain water become poisened. As demonstrated by acid lakes in Scandinavia and NE America, the acidification can reach the hydrosphere.

The problem now is how this poisening of the ecosphere affects the living beings: microorganisms, plants, animals, man. One has to expect effects due to accumulation of potential toxic substances in organisms or in their soil and water environment, and direct effects of airborn toxins like SO_2 and oxidants to plants, especially to leaves. It is easily to recognize that the interaction of direct and indirect effects, together with the natural stress and the stress caused by other activities of man (e.g. biomass utilization, use of biocids) should result in a great variety of growth disturbances and injuries. The acid environment should lead to wounds at the protection layers covering the outer surfaces of shoots and roots of plants. This should allow the entrance of dissolved toxins like heavy metal ions and of plant pathogens in the plant body. Root injury should lower the ability of the plant to react on all the various stresses. It will need many years of research, may be decades, before plant physiologists, plant pathologists and biochemists have checked all possible effects and their interaction.

Literature

Ellenberg, H. (ed) (1971): Integrated Experimental Ecology. Ecol.Studies (Springer Verlag) 2, 9-15

Matzner, E. et al. (1982): Elementflüsse in Waldökosystemen im Solling - Datendokumentation. Göttinger Bodenkdl.Ber. 71, 267 p.

Mayer, R. a. Ulrich, B. (1974): Conclusions on the filtering action of forests from ecosystems analysis. Ecol.Plant 9, 157-168

Mayer, R. (1981): Natürliche und anthropogene Komponenten des Schwermetall-Haushalts von Waldökosystemen. Göttinger Bodenkdl.Ber., 70, 1-292

Mayer, R. (1983): Interaction of forest canopies with atmospheric constituents: Aluminum and heavy metals. In B. Ulrich and J. Pankrath (eds): Accumulating Air Pollutants in Forest Ecosystems. D. Reidel Publ. Co., p. 47.

Ulrich, B., R. Mayer und P.K. Khanna (1979): Deposition von Luftverunreinigungen und ihre Auswirkungen in Waldökosystemen im Solling. Sauerländer Verlag Frankfurt

Ulrich, B. (1983a): Interaction of forest canopies with atmospheric constituents: SO_2, alcali and earth alcali cations and chloride. See MAYER 1983

Ulrich, B. (1983b): A concept of forest ecosystem stability and of acid deposition as driving force for destabilization. See MAYER 1983

REMARKS ON THE PAPER BY PROF. ULRICH
ON "EFFECTS OF ACID DEPOSITION"

G. Zimmermeyer
Gesamtverband des Deutschen Steinkohlenbergbaus
Essen

In his paper, Prof. Ulrich presented forest and tree damage in
areas which were the object of his studies. He concluded that
acid deposition (wet and dry) and other deposition of air pollutants might be responsible for the damage observed. As he
found damage exclusively in higher altitudes, he concluded
that the effects might be caused both by direct influence (via
deposition of fog and mist with the inherent air pollutants on
leaves and needles) and indirect influence (via deposition on
soil). He restricted his measurements mainly to the pathway
via the soil. Other experts in contrast to Prof. Ulrich stress
the opinion that the effects are likely to originate from direct influences. In this respect, the studies involving soil
and interception measurements reported by Prof. Ulrich are an
inadequate restriction and should be accompanied by measurements of pollution concentrations.

As no other expert in biological science was present, it
must be remarked that it has been impossible so far to furnish
proof of a far-reaching or only partial relationship between
the deposition of air pollutants and the forest damage observed. Other potential causes, such as climate, insect-pest
infestation, and changes in forestry patterns, are to be discussed, too. To date, any acceptable demonstration justifying
the exclusion of these parameters is outstanding. The demand
for an extension of these investigations with respect to
other possible causes and for a better coordination of the
different scientific fields involved was brought forward. In
drawing overall conclusions, experts in one field should more
intensely try to get expertise in other fields, too.

Far-reaching conclusions calling for further reductions
of SO_2-emissions to reduce forest damage remain unjustified.

ANSWER TO THE REMARKS OF MR. ZIMMERMEYER

B. ULRICH

The measurement of deposition gives information about the load to which
the ecosystem is subjected. The balances discussed allow to judge the
carrying capacity of the ecosystem; this judgement can be supplemented by
the knowledge of other state variables of the ecosystem as given elsewhere.
Knowledge of the load and of the carrying capacity allows conclusions
about consequences. Our investigations are directed to detect effects of
the different stress factors acting on the ecosystem, including climate,
biomass utilization and choice of tree species. These investigations have
already given considerable information about the interaction of the
various stress factors, including air pollutants. This information has
been presented in several papers (e.g. ULRICH et al. 1979, ULRICH 1983a
and b). Thus we have done precisly what ZIMMERMEYER demands but he seems
not to recognize this. The only environmental factor for forests which
has been changed is the "chemical climate" by air pollution. There is
therefore no doubt that this change is the driving force for a development
in the ecosphere which is characterized not only by tree and forest
die-back, but also by the acidification of waters and by the disappearence
of species at an increasing rate. The data about load, carrying capacity
and visible damage are more than enough to claim a rapid and considerable
reduction of air pollution to avoid a possible ecological catastrophe.
The facts mentioned indicate that the catastrophe may be already in
its initial stage - a century after the pollution of the ecosphere has
been started.

CHARACTERIZATION OF ACIDITY IN NATURAL WATERS

C. BROSSET
Swedish Water and Air Pollution Research Institute
P.O. Box 5207, S-402 24 Gothenburg

Summary

It has been stressed that a pH-value alone provides a poor characterization of the acid properties of natural waters.
The necessary information is usually obtained if the contribution
to the hydrogen ion concentration from different chemical individ-
uals and their identities are established. Two types of natural
waters have been discussed. The first one consists of the water soluble
part of airborne particles and precipitation. These natural waters,
when in internal and external equilibrium, may in many cases be described by the phase diagram for the system $H_2SO_4-H_2O-NH_3-HNO_3$.
This diagram is discussed.
The other type consists of lake waters. Their acidic properties may
be described by the total concentration of carbonic acid, strong
base, weak organic acids and strong (anthropogenic) acid. Possible
methods to determine these concentrations are presented.

1. Introduction

The acid properties of natural waters are usually defined by pH-
values. Sometimes these values are obtained by measurement *in situ*. In
other cases the pH is measured after the sample has been brought to the
laboratory. Normally these two types of measurement give rather different
results. It is also commonly recognized that a pH-value alone gives very
incomplete information on the acid properties of a natural water.

Better information may, however, be obtained if the measurements are
focused on those parameters which determine the relevant pH and not only
on the pH-value itself.

In a natural water these parameters consist of a number of chemical
groups acting as hydrogen ion donors (acids) or acceptors (bases). A complete characterization of the system implies knowledge of their identities, relevant physico-chemical properties and concentrations. Such comprehensive information is usually difficult to obtain.

Fortunately, however, the number of these parameters (acid-base pairs)
which seriously contribute to a pH-value is often rather limited within
a limited pH-range. As mainly anthropogenic acidification usually implies
a decrease of pH from 6-7 to 3-5 the last range is of special interest.

For that reason this paper will deal with some of the possibilities
we presently have of establishing in natural waters, those parameters
which are most responsible for pH being decreased in such a way.

General principles

For the sake of simplicity the following treatment of the subject is limited to monoprotic systems (with the (formal) exception of H_2SO_4 and H_2CO_3) containing two types of acid-base pairs:

$$HA \rightleftharpoons H^+ + A^-$$
$$BH^+ \rightleftharpoons H^+ + B \qquad\qquad \ldots\ldots \ (1)$$

The uncharged groups may be volatile or nonvolatile. The indices T, g, aq, and add. used below refer to total, gas phase, water phase, and added, respectively. [ ] means concentration.

As is well known, the acid-base pairs mentioned participate in the following equilibria:

HA is nonvolatile

$$[HA]_T = [HA]_{aq} + [A^-] \qquad\qquad \ldots\ldots \ (2)$$

$$\frac{[H^+] \cdot [A^-]}{[HA]_{aq}} = k_{HA}$$

Hence $[A^-] = \dfrac{k_{HA} \cdot [HA]_T}{[H^+] + k_{HA}}$ $\qquad\qquad \ldots\ldots \ (3)$

HA is volatile

$$\frac{[HA]_g}{[HA]_{aq}} = k'_{HA} \qquad\qquad \ldots\ldots \ (4)$$

Hence $[A^-] = \dfrac{k_{HA}}{k'_{HA}} \cdot \dfrac{[HA]_g}{[H^+]}$ $\qquad\qquad \ldots\ldots \ (5)$

B is nonvolatile

$$[BH^+]_T = [BH^+] + [B]_{aq} \qquad\qquad \ldots\ldots \ (6)$$

$$\frac{[H^+] \cdot [B]_{aq}}{[BH^+]} = k_{BH^+}$$

Hence $[BH^+] = \dfrac{[H^+] \cdot [BH^+]_T}{[H^+] + k_{BH^+}}$ $\qquad\qquad \ldots\ldots \ (7)$

B is volatile

$$\frac{[B]_g}{[B]_{aq}} = k'_B \qquad \qquad \dots (8)$$

Hence $[BH^+] = \dfrac{1}{k_{BH^+} \cdot k'_B} \cdot [H^+] \cdot [B]_g \qquad \dots (9)$

Addition of the acids HA_i and the bases B_i, nonvolatile or volatile, to water, generates in it the following hydrogen ion concentration:

$$[H^+] = \Sigma\,[A_i] - \Sigma\,[B_iH^+] + [OH^-] \qquad \dots (10)$$

For the $[H^+]$ -range $10^{-3} - 10^{-5}$ mole 1^{-1} three cases have to be discussed.

1) In eq. (2) $[HA]_T \cong [A^-]$

and in eq. (6) $[BH^+]_T \cong [BH^+]$

Such acids and bases obviously behave here as strong, and will be denominated HA_S and B_S, respectively.

2) In eq. (2) $[HA]_{aq}$ and $[A^-]$

and in eq. (6) $[BH^+]$ and $[B]_{aq}$

are of the same order of magnitude, respectively. Such acids and bases are weak. They here contribute to the value of $[H^+]$.

3) In eq. (2) $[HA]_T \cong [HA]_{aq}$

and in eq. (6) $[BH^+]_T \cong [B]_{aq}$

Such acids and bases are very weak. Here they do not contribute to the value of $[H^+]$.

From this discussion follows that it may be practical to separate the terms in equation (10) in such a way as to express the contributions to $[H^+]$ from acids or bases, which in the relevant $[H^+]$-range will act as strong or as weak. Using equations (3) and (7) this will give the following expression:

$$[H^+] = \Sigma\,[HA_S] - \Sigma\,[B_S] + \Sigma\,\frac{k_{HA_i} \cdot [HA_i]_T}{[H^+] + k_{HA_i}} - \Sigma\,\frac{[H^+] \cdot [BH^+]_T}{[H^+] + k_{BH^+}} + \frac{k_w}{[H^+]} \quad \dots(11)$$

In this equation the respective contributions are expressed in total concentrations. The terms corresponding to weak (monoprotic) components also contain the respective dissociation constants (and the ion product k_w for water).

In practical work evaluation of all these terms, if at all possible, implies very difficult measurements. The problem is in reality somewhat simplified as the only weak base, so far known of importance, in this connection is ammonia. In the relevant $[H^+]$-range, however, ammonia is present in water almost entirely as ammonium ion ($pk_{NH_4^+} = 9.25$).

Consequently, it will be taken care of by the term $\Sigma \ [B_s]$.
(It should be noticed that according to equation (9)
$[NH_4^+]$ will depend both on $[H^+]$ and $[NH_3]_g$).

The approximations mentioned will now reduce equation (11) as follows:

$$[H^+] = \Sigma \ [HA_s] - \Sigma \ [B_s] + \Sigma \ \frac{k_{HA_i} \cdot [HA_i]}{[H^+] + k_{HA_i}} T \qquad \ldots (12)$$

Accordingly, when investigating the acidity of a natural water, the first step will be to establish if weak acids are present. How this may be done is described in next section.

2. Establishing the presence of weak acids in a natural water

The method presented here is based on Gran's plot. (1)
Suppose a natural water sample is titrated with NaOH in such a way as to keep the volume of the sample (almost) constant. The NaOH-solution is added in equal doses and $[H^+]$ is measured after every addition. $[H^+]$ is finally plotted against $[OH^-]_{add}$ in a diagram.
As has been shown by the author (2) the slope (m) of the curve obtained in this way is given by the following expression:

$$\frac{d \ [H^+]}{d \ [OH^-]_{add}} = m = - \frac{1}{1 + \Sigma \ [HA_i]_T \cdot \alpha_i} \qquad \ldots (13)$$

For monoprotic acids we have

$$\alpha = \frac{1}{k_{HA} \left(1 + \frac{[H^+]}{k_{HA}} \right)^2} \qquad \ldots (14)$$

Hence for a strong acid $\alpha = 0$ and $m = -1$

In the presence of weak acids $m > -1$ as α is positive.
Obviously, the contribution from a weak acid to the slope m will depend on $[H^+]$, $[HA]_T$ and k_{HA}.

The variation of m with these parameters within relevant concentration intervals is given in Table 1 (3). From this table is seen which combinations of $[HA]_T$ and k_{HA} will perceptibly change the value of m in the $[H^+]$ -range, $10^{-3} - 10^{-5}$ mole l^{-1}. In other words, if during a titration of a water sample a change in m is observed, say in the $[H^+]$ - interval $10^{-3} - 10^{-5}$ mole l^{-1}, Table 1 gives some information about the weak acids possibly present and their concentration range.

It should be observed that if $[H^+] \cdot k_{HA}^{-1} \gg 1$, the acid HA is hardly dissociated at all and does not contribute to m. If $[H^+] \cdot k_{HA}^{-1} \ll 1$, the acid HA is almost completely dissociated and contributes to m with a constant value.

Linear extrapolition of the curve from the region where $\alpha \cong 0$ (and $m = -1$) to the abscissa obviously gives the concentration of the strong acid in the sample. This procedure is the original Gran's titration

Table 1

Values of $-m$ within $[H^+]$ interval $10^{-2} - 10^{-7}$ mole l^{-1} for various $[HA]_T$ and k_{HA}. The interval of $[H^+]$ value containing great changes of $-m$ is underlined

k_{HA}	$[HA]_T$ mole l^{-1}	$[H^+]=$ 10^{-2}	10^{-3}	10^{-4}	10^{-5}	10^{-6}	5×10^{-7}	10^{-7} mole l^{-1}
10^{-3}	10^{-2}	0.92	0.29	0.11	0.09	0.09	0.09	0.09
	10^{-3}	0.99	0.80	0.55	0.51	0.51	0.50	0.50
	10^{-4}	1.00	0.98	0.92	0.91	0.91	0.91	0.91
	10^{-5}	1.00	1.00	0.99	0.99	0.99	0.99	0.99
10^{-4}	10^{-2}	0.99	0.55	0.04	0.01	0.01	0.01	0.01
	10^{-3}	1.00	0.92	0.29	0.11	0.09	0.09	0.09
	10^{-4}	1.00	0.99	0.80	0.55	0.51	0.50	0.50
	10^{-5}	1.00	1.00	0.98	0.92	0.91	0.91	0.91
10^{-5}	10^{-2}	1.00	0.91	0.11	0.00	0.00	0.00	0.00
	10^{-3}	1.00	0.99	0.55	0.04	0.01	0.01	0.01
	10^{-4}	1.00	1.00	0.92	0.28	0.11	0.10	0.09
	10^{-5}	1.00	1.00	0.99	0.80	0.55	0.52	0.51
	10^{-6}	1.00	1.00	1.00	0.98	0.92	0.92	0.91
10^{-6}	10^{-2}	1.00	1.00	0.51	0.01	0.00	0.00	0.00
	10^{-3}	1.00	1.00	0.91	0.11	0.00	0.00	0.00
	10^{-4}	1.00	1.00	0.99	0.55	0.04	0.01	0.01
	10^{-5}	1.00	1.00	1.00	0.92	0.29	0.18	0.11
	10^{-6}	1.00	1.00	1.00	0.99	0.80	0.69	0.55
10^{-7}	10^{-2}	1.00	1.00	0.91	0.09	0.00	0.00	0.00
	10^{-3}	1.00	1.00	0.99	0.51	0.01	0.00	0.00
	10^{-4}	1.00	1.00	1.00	0.91	0.11	0.03	0.00
	10^{-5}	1.00	1.00	1.00	0.99	0.55	0.27	0.04
	10^{-6}	1.00	1.00	1.00	1.00	0.92	0.78	0.29
10^{-8}	10^{-2}	1.00	1.00	0.99	0.50	0.10	0.00	0.00
	10^{-3}	1.00	1.00	1.00	0.91	0.09	0.02	0.00
	10^{-4}	1.00	1.00	1.00	0.99	0.51	0.21	0.01
	10^{-5}	1.00	1.00	1.00	1.00	0.91	0.72	0.11
	10^{-6}	1.00	1.00	1.00	1.00	0.99	0.96	0.55
10^{-9}	10^{-2}	1.00	1.00	1.00	0.91	0.09	0.02	0.00
	10^{-3}	1.00	1.00	1.00	0.99	0.50	0.02	0.01
	10^{-4}	1.00	1.00	1.00	1.00	0.91	0.71	0.09
	10^{-5}	1.00	1.00	1.00	1.00	0.99	0.96	0.51
10^{-10}	10^{-2}	1.00	1.00	1.00	0.99	0.50	0.20	0.01
	10^{-3}	1.00	1.00	1.00	1.00	0.91	0.71	0.09
	10^{-4}	1.00	1.00	1.00	1.00	0.99	0.96	0.50
	10^{-5}	1.00	1.00	1.00	1.00	1.00	1.00	0.91

method (1). There is, however, still more information to be gained from such titration plots. Examples will begiven below.

3. Experimental

According to the above discussion the characterization of acidity in natural waters involves two types of measurement.

The first consists in the usual analysis of anion and cation concentrations.

The second is the titration procedure discussed in the last section. Details concerning this method are presented elsewhere (4). However, it is important to remember that all data must be obtained in terms of concentrations. They should correspond to points within the relevant $[H^+]$ -range ($10^{-3} - 10^{-5}$ mole l^{-1}). This is achieved by addition of KCl and $HClO_4$ to the sample before titration, giving it a concentration of 0.10 mole l^{-1} KCl and a $[H^+]>10^{-4}$ mole l^{-1}. Next CO_2 is removed by bubbling N_2 through the solution to facilitate the determination of other weak acids if present.

Finally, potentiometric titration is performed by addition of 0.0100 mole l^{-1} NaOH by means of a microburette. The electrode system is calibrated using solutions with known concentrations of $HClO_4$ in a 0.10 KCl-medium.

4. Characterization of the acidity of airborne particles and precipitation

The common features of leaching solutions of particles and precipitation are rather similar, at least if samples have been taken in remote areas and far from sea (sea salt concentration negligible).

After removal of CO_2, titration of such samples in the $[H^+]$ -range $10^{-3} - 10^{-5}$ mole l^{-1} usually gives $m = -1$. Further, roughly the following ion balance is prevailing

$$2\ [SO_4^{2-}] + [NO_3^-] = [H^+] + [NH_4^+]$$

which means that

$$[H^+] = 2\ [SO_4^{2-}] + [NO_3^-] - [NH_4^+]$$

Using equations (5) and (9) gives at equilibrium with the gas phase

$$[H^+] = 2\ [SO_4^{2-}] + \frac{[HNO_3]_g}{k'_{HNO_3} \cdot k_{HNO_3} \cdot [H^+]} - \frac{[NH_3]_g \cdot [H^+]}{k'_{NH_3} \cdot k_{NH_4^+}}$$

This interpretation is in fact based on the rather likely assumption that particles and raindrops originally acquired their acid properties by incorporating primarily formed H_2SO_4 (SO_2-oxidation). Subsequently, the liquid phase has equilibrated (or almost so) with respect to water vapour, ammonia and nitric acid in the air. In this case the composition of the solutions formed may be described by the relevant parts of the phase diagram for the system

$$H_2SO_4 - H_2O - NH_3 - HNO_3$$

This phase diagram in terms of a computer program is now being developed by R. Whitby and Y.H. Lee. (5)

But already in 1979 part of this diagram (H_2SO_4-H_2O-NH_3, acid part, $25^\circ C$) was presented by Y.H. Lee and the author (6) and, including some new points of view, by the author (7). However, a short presentation and discussion of this diagram are in order here. Fig. 1 shows the diagram.

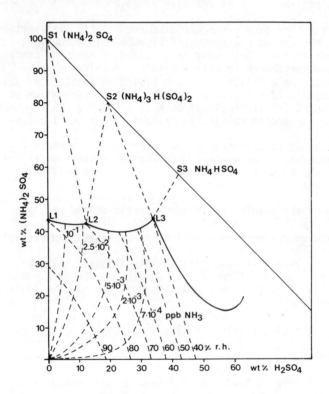

Figure 1. Phase diagram for the system $(NH_4)_2SO_4$ - H_2SO_4-H_2O at $25^\circ C$.

The choise of components ($(NH_4)_2SO_4$, H_2SO_4 and H_2O) has been proposed by Tang et al. (8). They have also determined the tielines for relative humidity (r.h.). The tielines for NH_3 have been calculated by Y.H. Lee (6). The solubility curve is based on data given by d'Ans (9). The solid phases have been identified in aerosol particles by the author (10).

The significance of the different points, lines and areas in the diagram is the following:

The line SI-S2-S3

The points SI, S2, S3 represent the solid phases $(NH_4)_2SO_4$, $(NH_4)_3H(SO_4)_2$ and NH_4HSO_4.

The line parts SI-S2 and S2-S3 correspond to mixtures of the phases SI and S2, and S2 and S3, respectively.

The curve LI-L2-L3

This is the solubility curve. The part LI-L2 represents solutions in equilibrium with SI. In the same way, the part L2-L3 represents solution in equilibrium with S2.

The solution at point L2 is in equilibrium with both SI and S2, and the solution at point L3 in equilibrium with both S2 and S3.

At points L2 and L3 there are, consequently, in this three-component system, four phases, which means one degree of freedom. At a chosen constant temperature both r.h. and $[NH_3]_g$ are thus given. At the curve parts denoted LI-L2 and L2-L3 there are three phases and thus two degrees of freedom. Consequently, at a chosen constant temperature, there is a further choice of, e.g., either r.h. or $[NH_3]_g$.

The area below the solubility curve

In this area, no solid phases are present, which means that the number of phases is two and so the number of degrees of freedom is three. At a chosen constant temperature there is a further choice of, e.g., both r.h. and $[NH_3]_g$.

However, such a chosen pair now defines the composition of the solution.

If the system also includes the component HNO_3, the composition of the solution will now in the same way be determined by r.h., $[NH_3]_g$ and $[HNO_3]_g$. The solution will then also contain NO_3^--ions.

The conclusion is now that $[H^+]$ in the liquid layer on airborne particles and in the precipitation will depend on (at least) r.h., $[NH_3]_g$ and $[HNO_3]_g$.

Especially r.h. and $[NH_3]_g$ are important. As is well known, r.h. may be very different at day and night. Low r.h. in the last case may push equilibrium conditions into a region where crystallisation should occur. Usually, however, a state of supersaturation will be attained.

The origin of $[NH_3]_g$ in the ground layer is usually the soil. For that reason it varies from region to region (e.g. high $[NH_3]_g$ over agricultural land, low over pinetree forest). It is also observed that in the liquid layer of airborne particles, the concentration ratio $[NH_4^+]/[H^+]$ will be altered as the air mass is moving over regions with different ammonia emissions.

5. Characterization of the acidity in lake water

Until now, only a few lake waters have been investigated in detail with respect to their acid properties. However, in all samples studied so far there have also been found weak acids other than H_2CO_3.

If in a water sample, after removal of CO_2, there is only one weak acid present, determination of its concentration and k_{HA}-value seems to be straightforward.

The titration described above will give a curve the slope of which is defined by the following expression (equations (13) and (14))

$$- m = \cfrac{1}{1 + \dfrac{[HA]_T}{k_{HA}} \cdot \dfrac{1}{\left(1 + \dfrac{[H^+]}{k_{HA}}\right)^2}} \qquad \ldots \ldots (15)$$

Measuring m and $[H^+]$ at two points on the curve will give the equations needed to solve $[HA]_T$ and k_{HA}.

A better method has been designed by Y.H. Lee (3). She rearranged equation (15) in the following way:

$$[H^+] = (\ [HA]_T \cdot k_{HA})^{+\frac{1}{2}} \cdot (-\ m^{-1} - 1)^{-\frac{1}{2}} \qquad \ldots \ldots (16)$$

$$\text{Let} \quad (-\ m^{-1} - 1)^{-\frac{1}{2}} = \xi$$

This will finally give:

$$[H^+] = (\ [HA]_T \cdot k_{HA})^{\frac{1}{2}} \cdot \xi - k_{HA} \qquad \ldots \ldots (17)$$

If $[H^+]$ is plotted against ξ, (calculated from measured values of m), a straight line will be obtained with the slope $([HA]_T \cdot k_{HA})^{\frac{1}{2}}$ and the intercept $-k_{HA}$. If Lee's plot does not give a straight line more than one weak acid is contributing to $[H^+]$ in the titration interval.

Using this method Lee has established the presence of what is probably the same weak acid in 6 different lake waters. In four of them there were no other weak acids (with the exception of H_2CO_3) observed.

From the good correlation between the colour of the samples and the concentrations of this acid it was concluded that the latter could represent a functional group(-COOH) in fulvic acid. Hence this acid is tentatively denominated $HA_{fulv.1}$ and its dissociation constant $k_{fulv.1}$. The result obtained by Lee is given in Table 2.

Table 2

Concentrations and dissociation constants of the acid $HA_{fulv.1}$

r = correlation coefficient of the plot

n = number of data pairs

Sample no.	$pk_{fulv.1}$	$[HA_{fulv.1}]_T \cdot 10^5$ mole$\cdot$l^{-1}	r	n
2	4.53	5.91	0.999	7
3	4.51	8.58	0.999	9
5	4.62	3.83	0.999	11
9	4.74	3.69	0.999	8

The remaining determination of strong acid or base in these lake water samples is straightforward. Remembering that the titration procedure here employed, involves addition of a known concentration of $HClO_4$ to the sample before titration, the observed $[H^+]$ will be composed in the following way:

$$[H^+]_{obs} = [H^+]_{HClO_4} + [HA]_S - [B]_S + \frac{k_{fulv.1} \cdot [HA_{fulv.1}]}{[H^+]_{obs} + k_{fulv.1}} T \quad \ldots (18)$$

or expressing the concentration of strong acid:

$$[H^+]_{str} = [H^+]_{HClO_4} + [HA]_S - [B]_S = [H^+]_{obs} - \frac{k_{fulv.1} \cdot [HA_{fulv.1}]}{[H^+]_{obs} + k_{fulv.1}} T \quad \overset{\ldots(19)}{}$$

Using data in Table 2 togheter with the respective titration data ($[H^+]_{obs}$ v. $[OH^-]_{add}$), values of $[H^+]_{str}$ could be calculated and plotted against $[OH^-]_{add}$. It was expected that $-m = 1$. The result is presented in Table 3 (from (11)).

Table 3

Calculated values for $[H^+]_{str}$, $[HA]_S$ and $[B]_S$,
and correlation coefficient (r) and slope (m)
for the plot $[H^+]_{str}$ v. $[OH^-]_{add}$

Sample no.	$-r$	$-m$	$[H^+]_{str}$	$[HClO_4]_T$	$[HA]_S$	$[B]_S$
					mole $l^{-1} \cdot 10^4$	
2	0.9999	1.01	2.083	1.692	0.391	
3	1.0000	0.97	2.396	2.353	0.043	
5	1.0000	0.98	2.137	2.681		0.544
9	1.0000	0.98	2.688	6.450		3.762

As is seen lake water samples 3 and 4 are heavily contaminated with (anthropogenic) strong acid, while sample 5 and 9 still have some alkalinity.

With the help of data hitherto obtained and using pH-measurements *in situ*, the original concentration of $[H_2CO_3]_T$ in the samples may be calculated. This involves, however, a difficult recalculation of pH to $[H^+]$. Direct methods are therefore recommended. It may, however, be mentioned that such approximative calculations indicate that in samples 2 and 3 CO_2 was present in heavy supersaturation in comparison with the concentration which would be expected if the sample had been in equilibrium with air.

Finally, it may be reported that for two lake waters Lee's plot was not quite linear (3).

Using a trial and error procedure, Lee was able to show that these samples probably contained $HA_{fulv.1}$ but also another acid, which is tentatively designated $HA_{fulv.2}$, as it may represent another functional group of fulvic acid (3). Her result is given in Table 4.

Table 4

Concentrations of $HA_{fulv.1}$ and $HA_{fulv.2}$ and their dissociation constants in two lake waters.

Sample no.	$pk_{fulv.1}$	$[HA_{fulv.1}]_T$ mole $1^{-1} \cdot 10^5$	$pk_{fulv.2}$	$[HA_{fulv.2}]_T$ mole $1^{-1} \cdot 10^5$
6	4.60-4.65	4.6-5.2	3.4-3.7	3.3-3.9
8	4.65-4.70	4.5	3.5-3.7	3.2-3.9

Evaluation of such systems would, of course, be much facilitated by the use of computer programs utilizing least square methods.

6. Conclusions

It has been shown that in the liquid film on airborne particles and in the precipitation, hydrogen ion concentration ($[H^+]$) may, in many cases be looked upon as a non-neutralized rest of H_2SO_4 and HNO_3. As the neutralizing agent usually is NH_3 from the air, the process of neutralization is here reversible and the degree of neutralization depends among other things on the concentration of ammonia in the air ($[NH_3]_g$).

If equilibrium in the air-water system is prevaliling, relevant concentrations in the liquid phase may be calculated if the total concentration of sulphate in the solution ($[SO_4^{2-}]_T$), the relative humidity (r.h.), and the concentrations of ammonia and nitric acid in the gas phase ($[NH_3]_g$ and $[HNO_3]_g$ resp.) are known.

On the other hand, measurement of concentrations both in the air and in the solution will give information as to how far the system is from equilibrium.

Further, it has been shown that the hydrogen ion concentration ($[H^+]$) in acidified lakes mainly depends on the concentrations of carbonic acid ($[H_2CO_3]_T$), strong base ($[B]_c$), organic weak acids ($[HA_i]$) and above all, on the concentration of deposited strong acid ($[HA]_c$). A possible way of determining the total concentrations of these compounds has been described.

It has been stressed that measurement of pH alone provides poor information about the acid properties of a natural water.

References:

1. Gran, G. 1952, Analyst 77, 661
2. Brosset, C. 1976, WASP 6, 259-275
3. Lee, Y.H. 1980, WASP 14, 287-298
4. Lee, Y.H., Brosset, C. 1978, WASP 10, 457-469
5. Private communication
6. Lee, Y.H., Brosset, C. 1979, WMO Symposium, Sofia,
   Bulgaria, 1-5 October. WMO - No. 538, Geneva,Switzerland
7. Brosset, C. 1979. ORNL Life Sciences Symposium on Potential
   Environmental and Health Effects of Atmospheric Sulfur
   Deposition. , Gatlinburg, Tennessee, October 14-18.
   IVL B -Report No. 516
8. Tang, J.N. et al. 1978, J. Aerosol Sci. 9, 505-511
9. D'Ans, J. 1913, Z. allg. anorg. Ch. 80, 235-245
10. Brosset, C. et al. 1975, Atm. Environ. Vol.9, 631-642
11. Brosset, C. 1980. WASP 14, 251-265

FORMATION OF ATMOSPHERIC ACIDITY

R.A. Cox
S.A. Penkett
Atomic Energy Research Establishment Harwell Didcot Oxon UK

Summary

Atmospheric oxidation is a natural process which is
largely driven by photochemistry. In the case of oxides
of nitrogen and sulphur emitted from the combustion of
fossil fuels, oxidation leads to the formation of strong
acids which cause acidification of rain in areas far
removed from their source. The mechanisms of the various
oxidation processes are dealt with in this paper. They
can occur in the gas phase, in liquid droplets or possibly
on the surface of some aerosol particles. They are very
complex and involve free radicals, such as the hydroxyl
radical, strong oxidant molecules, such as ozone and
hydrogen peroxide and transition metal catalysts.

Much progress in the understanding of these phenomena was
made under the auspices of the COST 61A program but many
outstanding problems remain. Some recent work in this
field is reviewed and a list of current research
priorities is appended. Emphasis is placed on the need to
understand seasonal phenomena, the role of droplet phase
chemistry and the interaction of fossil fuel emissions
with the emissions from motor vehicles.

INTRODUCTION

The understanding of the problem of acid deposition
requires knowledge of the relationship between the distribution
in time and space for the emission of the pollutants giving
rise to atmospheric acidity, and that for deposition of those
pollutants back to the earth's surface. The situation is
complicated because both the primary pollutants SO_2, NO, HCl,
and the secondary pollutants, NO_2, nitric acid, sulphuric acid
and sulphate and nitrate aerosols, are involved.
The major primary pollutants of relevance to the acid
deposition problem are sulphur dioxide, SO_2, nitric oxide, NO
and hydrogen chloride, HCl. Smaller amounts of the higher
oxides SO_3 and NO_2 are also released. Combustion of fossil
fuels in stationary and mobile energy generation plant is by
far the most important source for these pollutants. After
release the gases disperse into the atmospheric boundary layer
from whence they are ultimately removed either by dry
deposition at the earths surface or by incorporation into the
precipitation elements. However since the atmosphere is an
oxidising environment, chemical oxidation of the primary
pollutants occurs as they are dispersed. In this process the

gaseous oxides SO_2 and NO are converted to low volatility oxidation products H_2SO_4 and HNO_3 which form aerosols and are very readily incorporated into the precipitation elements. These oxidation products are strong acids and their presence largely detemines the pH of cloud droplets, rain and snow. HCl is itself a strong acid and can contribute to precipitation acidity by solution into cloud and rain droplets without chemical oxidation.

A knowledge of the rates and pathways of these oxidation processes is necessary to describe the fate of the pollutants from source to sink. Early attempts to address this problem concentrated mainly on the oxidation of sulphur dioxide. SO_2 is a water soluble gas and it was well known that in O_2 saturated solutions oxidation of the various S^{iv} species to sulphates (S^{vi}) can occur rapidly, particularly in the presence of certain metal-ion catalysts. Consequently this 'droplet phase' process was considered to be responsible for the acid sulphates present in rain.[1] However SO_2 emissions were often associated with dust and smoke too, and heterogeneous catalysis of the exothermic oxidation of SO_2 to SO_3, with subsequent hydration of SO_3 was also thought by some to be responsible for the

$$\begin{array}{c} \text{surface} \\ SO_2 + O_2 \quad \text{-->} \quad SO_3 \\ SO_3 + H_2O \quad \text{-->} \quad H_2SO_4 \end{array}$$

formation of highly acidic fogs in industrial and urban atmospheres.

Homogeneous reaction of SO_2 with atmospheric oxidising molecules, O_2 and O_3 is too slow to be significant, but early studies in California identified the phenomenon of photo-chemical air pollution, where oxidation of certain pollutants was driven by sunlight in processes involving photochemically produced free radical intermediate species.[2] Some evidence for photochemical oxidation of SO_2 to H_2SO_4 was found during these earlier studies but more important was the clear recognition of the key role of the nitrogen oxides in the photochemistry and the characterisation of some of the key reactions involved in the conversion of NO to nitric acid, through NO_2. More recently better characterisation of the role of photochemical reactions in SO_2 oxidation has been achieved in Europe and the USA.

Advances in knowledge in the field of atmospheric chemistry in general in recent years has resulted mainly from the study of the life-cycle of trace constituents in the atmosphere on a global or regional scale. A fundamental role for driven chemistry the oxidation and removal of trace gases in both the natural and polluted atmosphere is now recognised, and a reasonably detailed description of the basic mechanisms of these photochemical reactions is now available.[3] The importance of photochemically derived oxidants in droplet phase chemistry has also been explored.[4] The key chemical factors resulting from absorption of solar radiation in an atmosphere containing O_2 and water vapour can be summarised as follows:

    (a)   A relatively high abundance of ozone resulting from photochemical production of O atoms from O_2 (in the upper atmosphere) and from NO_2 (in the lower atmosphere). O_3 is then formed in the reaction

$$O + O_2(+M) \longrightarrow O_3(+M) \quad M = N_2, O_2, H_2O \text{ or } CO_2$$

(b) The presence during daytime of low steady state concentrations of OH radicals and other oxidising species e.g. HO_2, H_2O_2, originating from the reaction with water vapour of the excited atomic oxygen photofragments from ozone photolysis in the middle UV region:

$$O_3 + h\nu \longrightarrow O(^1D) + O_2.$$
$$O(^1D) + H_2O \longrightarrow 2\,OH$$

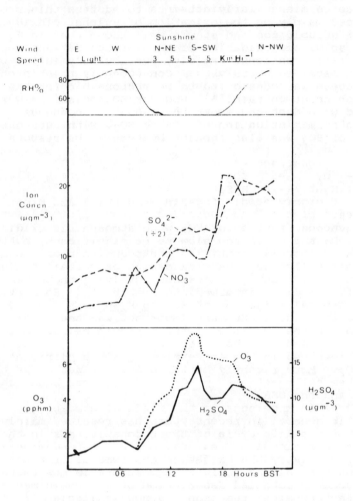

Figure 1   **Photochemical episode at Silwood Park, U.K.**

(c)   The strong coupling involved in the oxidation of many
      different atmospheric trace-gases, and consequently a
      strong dependence of overall reaction rates on trace
      gas composition.
(d)   The provision of a source of oxidising agents (e.g.
      O_3, H_2O_2), available for cloud-water chemistry, the
      cloud droplets providing a medium where ionic species
      derived from soluble trace gases may be oxidised.
The following sections describe in some detail how these
oxidising systems act on the primary pollutants leading to the
formation of atmospheric acidity.

1.   HOMOGENEOUS OXIDATION REACTIONS
1.1   Sulphur Dioxide
      Some of the earliest evidence for photochemical oxidation
of sulphur dioxide in the atmosphere comes from observation of
the relationships between ozone, sulphate and sulphuric acid
aerosol in the atmosphere in Southern England.[5]   Fig.1 shows
the variation of these components with time during a
photochemical pollution episode at Silwood Park, UK, during
September 1972.   The characteristic diurnal variation in O_3 was
closely followed by the relatively high H_2SO_4 levels and there
was a steady increase in aerosol sulphate and nitrate ion
during the day.   These changes occurred during conditions of
bright sunlight, light variable winds and low relative
humidity.   Cloud related processes were unlikely to be a medium
for formation of the oxidised sulphur (and nitrogen) species in
this case.

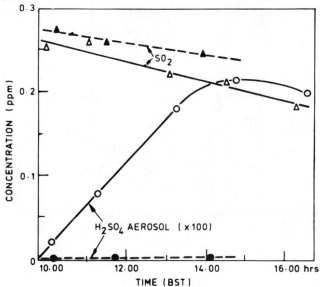

Figure 2   Formation of H_2SO_4 aerosol during
                  irradiation of SO_2-air mixtures in natural
                  sunlight. Filled points show results with
                  no exposure to sunlight.

Further important evidence for photochemical oxidation of SO_2 in sunlight comes from laboratory studies in which natural air containing added trace amounts of SO_2 was exposed to sunlight in transparent inert plastic-film bags.[6,7] Fig.2 illustrates rather clearly the steady conversion of SO_2 to H_2SO_4 aerosol in these experiments carried out on COST 61a.[6]

Control experiments in which the mixtures were kept in the laboratory showed essentially zero conversions. It was found in a series of these experiments that the rate of SO_2 oxidation was a complex function of the trace gas composition of the air mixtures (e.g. NO,NO_2, hydrocarbon levels) as well as being dependant on weather and seasonal factors such as solar intensity and temperature. For this reason quantitative evaluation of SO_2 oxidation rates requires detailed considera- tion of the mechanism by which oxidation occurs, and work on this problem in recent years has been largely in this direction.

The current picture of homogeneous oxidation mechanisms of SO_2 and other sulphur compounds is summarised schematically in Fig.3.

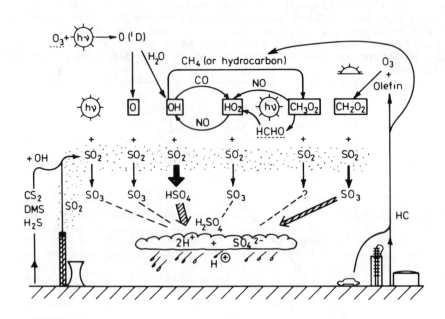

Figure 3   Homogeneous Oxidation of Atmospheric SO_2

A full discussion of all processes has been given by Calvert et al.[8] SO_2 released from a combustion source encounters a variety of gaseous oxidising entities, when dispersed in the free atmosphere. The rate of SO_2 oxidation by a particular component is given by

$$- \frac{d\,[SO_2]}{dt} = k\,[X][SO_2]$$

where k is the rate coefficient for reaction and X the concentration of the oxidising component, assumed to be at steady state. The reciprocal of the pseudo first order rate coefficient $(k\,[X])^{-1}$ gives the lifetime of SO_2 with respect to this process. The rate coefficients are determined in laboratory experiments and the steady state concentration of the various components are obtained from model calculations of the atmospheric chemical cycles.[3] Knowledge of these quantities allows the oxidation rate to be determined. The basic skeletal structure of the chemistry leading to oxidising species is also outlined in figure 3. The production of OH radicals from O_3 photodissociation and the reaction of $O(^1D_2)$ with H_2O, is followed by a series of fast elementary steps in which OH and HO_2 radicals are interconverted e.g.

$$OH + CO \quad \longrightarrow \quad CO_2 + H$$
$$H + O_2(+M) \quad \longrightarrow \quad HO_2(+M)$$
$$HO_2 + NO \quad \longrightarrow \quad NO_2 + OH$$

Radicals are removed by reaction with each other in terminating reactions, which may lead to oxidising products e.g.

$$HO_2 + HO_2 \quad \longrightarrow \quad H_2O_2$$

or to acidic gases e.g.

$$OH + NO_2(+M) \quad \longrightarrow \quad HNO_3(+M)$$

The total concentration of radicals depends on the relative rates of photochemical production and termination, whilst the rates of the interconversion reactions determine the relative abundance of individual radicals. The impact of other pollutant species, particularly hydrocarbons and NO_x, is indicated. Hydrocarbons are oxidised through peroxy radicals, RO_2, to aldehydes including formaldehyde, which are photo-chemically active. Photodissociation of aldehydes provides additional input of radicals in the sunlit atmosphere, thereby enhancing photochemical oxidation rates of all species. Nitric oxide plays a fundamental role in converting the unreactive RO_2 and HO_2 species to reactive RO and OH radicals, whereas NO_2 removes OH radicals as HNO_3.

There is now an adequate data base[3,9] for the kinetics of the elementary thermal and photochemical reactions to predict the average steady state radical concentrations of radical species for typical sunlit unpolluted atmospheres. Under these conditions, SO_2 oxidation is dominated by attack by OH radicals[8]

$$OH + SO_2(+M) \quad = \quad HSO_4(+M)$$

The detailed fate of the HSO_3 radical has not been fully elucidated but under simulated atmospheric conditions very efficient conversion of the products of this reaction to H_2SO_4 aerosol has been observed.[6] It is still not known whether the overall reaction removes OH radicals i.e. acts as a terminating reaction. This could be important in plume chemistry where the relatively high SO_2 levels prevailing could perturb the OH steady state concentration.

In polluted air OH-initiated oxidation is still dominant but other reactions may, under some circumstances, be significant. For example the thermal reaction between O_3 and olefinic hydrocarbons leads to reactive intermediates. These species have not been fully characterised but laboratory studies clearly show they can efficiently oxidise SO_2 to SO_3.[10,11]

Finally the direct photo-oxidation of SO_2 has to be carefully considered. SO_2 molecules absorb UV radiation quite strongly in the near UV region. Although there is insufficient energy in this radiation to break the OS - O bond, a number of reactions of electronically excited SO_2 molecules have been characterised. On the basis of what is known about the excited molecules the reactions occurring in ambient air do not lead to net oxidation of SO_2 at an appreciable rate, since the excited molecules are rapidly quenched to the ground state by N_2, O_2 and H_2O.[12]

None of the homogeneous oxidation reactions of SO_2 lead to very rapid oxidation rates under ambient atmospheric conditions. Under field conditions the most rapid observed oxidation rates which could be ascribed to homogeneous, photochemically initiated reactions are of the order of 3-4% hr^{-1}.[13] However the reactions continue steadily during daylight hours in the absence of meteorological pertubations, and significant overall conversion of SO_2 to acid sulphate can result. On the basis of model calculations,[14] typical life-times for SO_2 with respect to OH-initiated oxidation in NW European summertime conditions are approximately 10 days with a maximum oxidation rate of the order of 2% hr^{-1} at full sunlight. Wintertime rates of homogeneous SO_2 photo oxidation under clear-sky conditions are expected to be a factor of 3 to 5 slower.

Realistic estimates of the average rates of the ozone-olefin induced SO_2 oxidation reaction are difficult because of the paucity of information on ambient levels of olefins, and the different reactivities of the various olefins. However, with the exception of special situations relating to large olefin sources, the SO_2 oxidation rate is not expected to exceed 0.1% hr^{-1}. Note that this mechanism does not require photochemical initiation and can occur in the dark. An upper limit rate of the order of 0.1% hr^{-1} has also been estimated[8,15] for direct photo oxidation of SO_2 involving the electronically excited SO_2 radicals and this process is therefore much less important than the reactions involving free radical attack.

The oxidation of other gaseous sulphur compounds, H_2S, mercaptans, dimethyl sulphide, CS_2, is also important for the atmospheric sulphur cycle. These compounds are all oxidised by homogeneous reactions involving OH radical attack[3] and SO_2 is, in all cases, an important oxidation product. The rate of oxidation of the organic sulphides is very rapid (life-times $\leqslant$ 1 day) but H_2S and CS_2 are oxidised more slowly. CS_2 is also oxidised by direct photo-oxidation reactions involving electronically excited CS_2^* molecules.[16] The overall life-time of CS_2 with respect to homogenous oxidation is of the order of 10 days.

1.2 Nitrogen Oxides

Although observational evidence for the oxidation in the
atmosphere of nitric oxide, NO, to nitrogen dioxide, NO$_2$ has
been widely reported in photochemical smog, in non-urban air
and in power plant plumes the demonstration of the relationship
between atmospheric NO$_2$ and nitric acid has only appeared
recently.[17] This has arisen from the difficulty of measuring
gaseous nitric acid and also the significant volatility of
nitrate aerosols, with consequent ambiguity in interpretation
of data.

There is now a considerable body of laboratory data for
the gas phase homogeneous reactions which control the
interconversion and oxidation of atmospheric nitrogen oxides,
and their influence on the chemistry of other atmospheric
constituents.[9] In addition to the commonly known oxides of
nitrogen NO and NO$_2$, reactions involving the higher oxides
nitrogen trioxide, NO$_3$, and nitrogen pentoxide, N$_2$O$_5$ as well as
nitrous acid HONO, nitric acid and peroxynitric acid are
important.

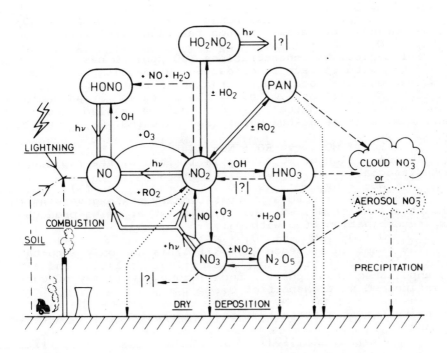

Figure 4   Chemical Transformations of Atmospheric NO$_x$

Interactions of NO_x with hydrocarbons occur both in urban and non-urban situations and so the reactions involving organic derivatives alkyl nitrites, alkyl nitrates and peroxynitrates, must also be considered. Especially important are the peroxyacylnitrates (PAN's) which observational data[18] show to be present in non-urban air as well as being important secondary pollutants in photochemical smog.

Figure 4 shows an illustration of the currently accepted picture of the cycle of gaseous nitrogen oxides through the lower atmosphere. The basic model is similar to that formulated originally by Robinson and Robbins in 1971[19] but has been revised somewhat in the light of recent observations related to sources and distribution of NO_x species and new chemical kinetic data.[20]

The main source of NO_x in the lower atmosphere is emission of NO at the earth's surface, either from soil processes or from man-made sources, mainly combustion, including biomass burning. Additional tropospheric sources are N_2 fixation by lightning and NH_3 oxidation, with some contribution from downward transport of stratospheric NO_x.[21] There is still considerable uncertainty about the magnitude, both relative and absolute, of these natural sources. It is probable, however, that the man made source is dominant in populated regions such as western Europe.

Oxidation of NO

Once in the atmosphere chemical oxidation of NO to NO_2 occurs rapidly, primarily through reaction with ozone:
$$O_3 + NO \longrightarrow NO_2 + O_2 \qquad (1)$$
For typical O_3 concentrations (30 ppb) NO has a life-time of $\approx$ 1 min with respect to oxidation by reaction (1). However if NO exceeds the local concentration of that of O_3, the reaction rate will be limited by the diffusion or advection of ozone into the air parcel. In daylight NO is also oxidised by photochemically generated peroxy radicals (HO, CH_3O_2, $CH_3CO.O_2$ etc),
$$RO_2 + NO \longrightarrow RO + NO_2 \qquad (2)$$
but net conversion of NO to NO_2 is limited by the rapid photodissociation of NO_2 which reforms NO:
$$NO_2 + h\nu \longrightarrow NO + O \qquad (3)$$
This gives rise to a photostationary state governing the concentrations of NO, NO_2 and O_3, the latter being produced by fast recombination of O with O_2:
$$O + O_2 + M \longrightarrow O_3 + M \qquad (4)$$
The reaction sequence (2)+(3)+(4) is the only important chemical source of ozone in photochemical pollution and also in the unpolluted troposphere. Local ozone levels may of course be influenced by transport of O_3.

The ratio of NO : NO_2 is determined only by the light intensity (as it affects k_3), the temperature (as it affects k_1) and $[O_3]$:
$$\frac{[NO_2]}{[NO]} = \frac{k_1 [O_3]}{k_3} \qquad (i)$$
The relaxation time for eq (i) is 1-2 min in sunlight.

A minor pathway for NO is reaction with OH to form nitrous acid (HONO). Photodissociation of HONO (typical photochemical life-time 10-20 mins) reforms NO and OH. HONO can be also

formed by a heterogeneous reaction of NO, NO_2 and water vapour.
Recent observations[22] confirm the occurrence of this reaction
in polluted air, where it can provide a significant source of OH
radicals.

Oxidation of NO_2

Oxidation of NO_2 mainly leads to formation of nitric acid.
In daylight this occurs by reaction with OH:
$$OH + NO_2 + M \quad --> \quad HONO_2 + M \qquad (5)$$
The rate of this reaction is dependent on the concentration
of photochemically generated OH which is predicted to be highly
variable; the life-time for NO_2 with respect to oxidation via
reaction (5) is of the order of 1 day in mid-latitude summer
time. Conversion of NO_2 to nitric acid can also occur through
reaction with O_3.
$$NO_2 + O_3 \quad --> \quad NO_3 + O_2 \qquad (6)$$
The initial product NO_3 forms nitrogen pentoxide in the
reversible reaction:
$$NO_3 + NO_2 + M \quad --> \quad N_2O_5 + M \qquad (7)$$
$$<--$$

N_2O_5 reacts with liquid water to form HNO_3. The efficiency
of this process is much reduced in daytime due to the alternate
removal processes for NO_3, i.e. reaction with NO to form NO_2:
$$NO_3 + NO \quad --> \quad 2NO_2 \qquad (9)$$
and photolysis to give NO or NO_2:
$$NO_3 + h\nu \quad --> \quad NO + O_2 \qquad (10a)$$
$$--> \quad NO_2 + O \qquad (10b)$$
Photolysis of NO_3 in these two reactions is very rapid
(half-life 10 s in sunlight) and this process dominates NO_3
loss in daylight, except when $[NO_x] > 3$ ppb or in the presence of
cloud droplets. The rate of oxidation of NO_2 by O_3 at nightime
is of the same magnitude as the daytime rate for OH reaction.
Formation of organic peroxy nitrates and peroxynitric acid
are also significant reaction pathways for atmospheric NO_2 e.g.
$$CH_3CO.O_2 + NO_2 + M \quad --> \quad CH_3COO_2NO_2 + M$$
$$<-- \qquad (PAN)$$
$$HO_2 + NO_2 \quad --> \quad HO_2NO_2$$
$$<--$$
Both these reactions are reversible and the extent to which
they provide a sink for NO_2 depends on the thermal stability of
the particular peroxy compound and the rates of the alternate
reactions it undergoes. In the lower atmosphere HO_2NO_2 only
plays a minor role since it is very unstable with respect to
thermal decomposition (half-life $\approx$ 10 s).[23] Peroxyacetyl-
nitrate (PAN) is more stable (half-life $\approx$ 1 hr)[24,25] and
consequently it is a more important 'NO_2 sink' and, in polluted
air, is a significant secondary pollutant. PAN is probably not a
significant factor for acidity in rain since it has limited
solubility and hydrolysis leads to nitrous acid, which is a weak
acid and is also volatile. Nevertheless the chemistry of PAN and
related compounds may be important in the transport of NO_x to
far-field sinks.[25]

1.3 Hydrogen Chloride

Since gaseous HCl is relatively stable towards oxidative
attack and is very soluble, the major fate of atmospheric HCl is
expected to be incorporation into the precipitation elements or
dry deposition. There is little observational data available on
this gas in order to confirm this picture, however.

The rate coefficient for reaction of OH with HCl is similar to that with SO_2

$$OH + HCl = H_2O + Cl$$

and homogeneous oxidation of HCl is therefore expected to occur at a similar rate in the sunlit atmosphere i.e. 1 to 2% hr^{-1} in summertime. Fig.5 shows an outline of the reactions expected to occur following the OH + HCl reaction. This scheme is based largely on what is known about the atmospheric chemistry of chlorine[9] which has received great attention in recent years in connection with the effect of chlorofluoromethanes on stratospheric ozone.[27]

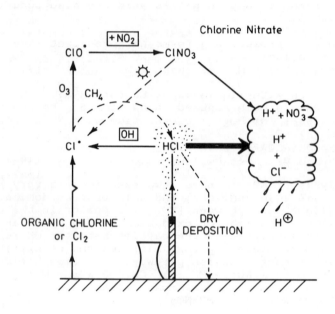

Figure 5   Chemical Transformations of HCl

Atomic chlorine can react by two paths of approximately equal rate in the lower troposphere. Reaction with atmospheric methane (or other hydrogen containing trace gas) simply reforms HCl again, giving no net oxidation of HCl. Reaction with ozone gives rise to the ClO radical, which is subsequently converted to the stable molecule chlorine nitrate, $ClONO_2$. Chlorine nitrate photodissociates only slowly to release Cl atoms, but it reacts rapidly in liquid water undergoing hydrolysis[28]

$$ClONO_2 + H_2O \longrightarrow HNO_3 + HOCl(= H^+ + ClO^-$$

This sequence therefore ultimately leads to the same strong acid input to precipitation elements as the major NO_2 and HCl pathways but in addition, the oxidising ion OCl^- is incorporated. More data is required to obtain a firmer picture regarding the behaviour of gaseous Cl-containing species in the troposphere.

2. LIQUID PHASE OXIDATION REACTIONS
2.1 General Introduction

Liquid phase oxidation can take place in the atmosphere in clouds, fogs and in rain droplets. It may also occur in hygroscopic aerosol droplets but the extent of liquid phase available here is very small. At any one time about one tenth of the atmosphere is occupied by cloud droplets, so there is plenty of scope for droplet reactions to occur. Fogs are a rather special form of cloud which occur very close to the ground where pollutant concentrations can be much higher; this is parti-cularly true of urban fogs which have presented a major pollution hazard in the past. Rain drops have a very short life in the atmosphere, probably not much longer than 2 minutes. They also have quite a small surface to volume ratio compared to cloud droplets and the combination of these two circumstances reduces their effectiveness as a reaction medium.

The liquid water content of many clouds is in the range of 1 gm m^{-3} of air, for fogs it is about an order of magnitude less and for aerosols probably 4 to 5 orders of magnitude less. At most the droplet reaction medium is only one millionth of the gas phase reaction medium and therefore only a few reactions will occur more efficiently within atmospheric droplets, clouds or otherwise. For those that do occur it is essential that the reactants are highly soluble, which generally means that extensive hydrolysis occurs, and reactions in droplets will be favoured if the hydrolysis products are more chemically reactive.

Sulphur dioxide is a very soluble gas and observations made over many years suggest that it is oxidised in fog droplets to give sulphuric acid. Coste and Courtier [29] pointed out in 1936 that substantial levels of H_2SO_4 were only observed in London on foggy days. Very recently Hoffman[30] and his co-workers at Caltech have made a similar observation in Los Angeles with regard to episodes of low visibility associated with high sulphate levels. These are always preceded by the formation of a fog containing an extensive droplet phase early in the morning. From about 1930 onwards investigations have been carried out on the role of catalytic substances promoting sulphuric acid production in fogs with the main ones being identified as iron and manganese.[31-42]

Another observation which has been interpreted as being indicative of the droplet phase oxidation of SO_2 is the omnipresence of ammonium sulphate in rainwater. Junge and Ryan[1] proposed that this was also the result of catalytic oxidation which was taking place in very remote areas and which was assisted by the presence of sufficient ammonia to neutralise the sulphuric acid formed. Since this proposal there has followed a succession of papers on the general theme of SO_2 oxidation assisted by ammonia but not requiring the presence of trace metal catalysts. The SO_2 in this case being simply oxidised by molecular oxygen which is present in all atmospheric water.

A most plausible scheme was put forward by Scott and Hobbs[43] who used van den Heuvel and Mason's[44] rate data to predict the rate at which ammonium sulphate would build up in cloud water. A chemically more elegant paper by McKay[45] used Fuller and Crist's[46] experimentally determined rate of sulphite oxidation by oxygen in solution to predict much faster rates of ammonium sulphate production than Scott and Hobbs.[43] These papers contained a lot of careful thought about the way in which sulphate formation should occur in rain clouds and McKay[45] in particular sought a rapid process that did not depend upon the presence of metal catalysts. These papers also assumed that the non catalysed droplet mechanism would explain the presence of ammonium sulphate aerosol which exists universally throughout the troposphere.

In 1972 Penkett[4] proposed an uncatalysed mechanism of SO_2 oxidation which involved ozone rather than oxygen as the oxidising agent. Experiments performed in 1973 in a fog chamber by Penkett and Garland[47] showed that ozone was much more efficient than oxygen as an oxidant for SO_2; this has subsequently been confirmed by other workers notably Erikson et al,[48], Larson and Harrison[49], Hegg and Hobbs[50].

During the period that this work was going on (1950 to 1970 approximately) a fundamental change was occurring in the pattern of emissions of pollutants. Emissions of black smoke from coal burning were declining such that the average concentration of smoke at a background site like Harwell dropped from 150 mg/m^3 in 1958 to 50 mg/m^3 in 1973 (Salmon, Atkins et al),[51] and it continues to decline today. Black smoke of course is the major source of potentially catalytic substances in the atmosphere of Western Europe and the North Eastern United States of America, and, as a consequence, the concentrations of species such as manganese and iron have declined at a similar rate.

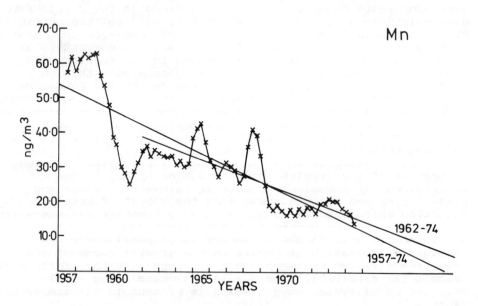

Figure 6   Decline in manganese concentration in the aerosol measured at Harwell U.K. over the years 1957 to 1974.

Figure 6 illustrates the magnitude of the decline for one
element, manganese, which is one of the most effective catalysts
for SO_2 oxidation in solution.   If catalytic oxidation were to be
a dominant mechanism for both producing sulphate in rainwater and
as an aerosol, a similar decline would have been expected for
sulphate.   In fact the European rain network shows no such
decline in the concentration of sulphate ion and the sulphate
aerosol concentration has increased since the mid-fifties at
Harwell[51] (Figure 7).   The sulphate data when divided by season
shows no trend in the winter and an upward trend close to 4% per
year in the summer.   Any solution phase mechanism responsible for
these observations is unlikely to involve metal catalysts.

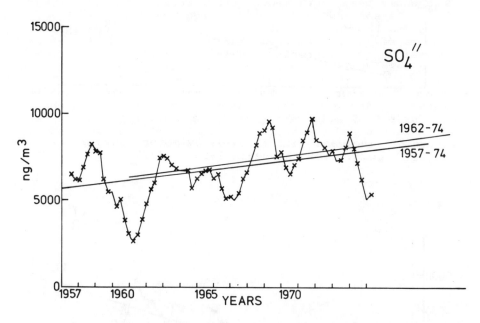

Figure 7   Increase in sulphate concentration in the
              aerosol measured at Harwell U.K. over the years
              1957 to 1974.

      Whilst black smoke emissions were declining, emission of
hydrocarbons and oxides of nitrogen from motor vehicles were
increasing.   Figure 8 shows the increase in fuel usage by motor
vehicles that took place in the U.K. through the same period
covered by the data shown in Figures 6 and 7.   Whereas there is
no data for the concentration of the reactant species (HC's and
NO_x), the nitrate aerosols trend shows a dramatic increase[37]
(Figure 9).   A similar increase for nitrate in rainwater has also
been observed both in Europe and the U.S.A.[52]   These
observations of increasing nitrate, and sulphate, particularly in
the summer months, strongly suggest the importance of an
underlying photochemical mechanism involving hydrocarbons and
nitrogen oxides emitted from motor vehicles.

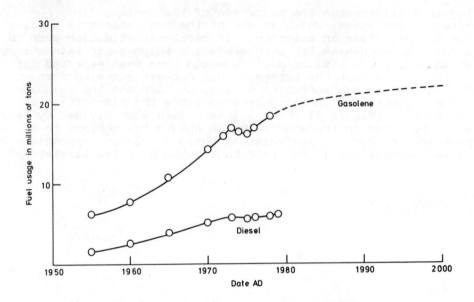

Figure 8   Increase in fuel usage by motor vehicles in the U.K. over the years 1955 to 1979

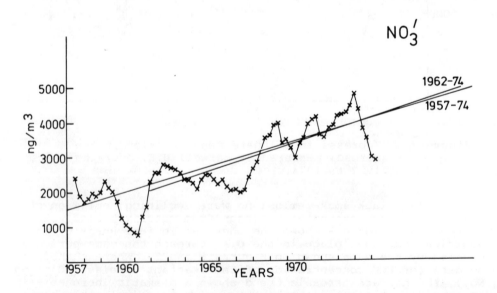

Figure 9   Increase in nitrate concentration in the aerosol measured at Harwell U.K. over the years 1957 to 1974

The importance of photochemistry in generating sulphate aerosols in the atmosphere by gas phase processes has been discussed above. Much of this work was performed under the COST program 61a which contained the original proposal that oxidation of SO_2 in droplets by hydrogen peroxide could be a most effective way of producing acid rain.[53] The COST 61a program stimulated a lot of interest in kinetic studies on liquid phase processes[53,54], some of which was reported at the Dubrovnik conference in 1978.[55]

2.2 Kinetic Studies Performed under the Auspices of COST 61a

In an attempt to assess the efficiency of the catalytic process as it might operate under atmospheric conditions, rate studies were performed using rainwater as a reaction medium.[54] This suggested that rates of SO_2 oxidation could be calculated using two simple equations involving the manganese concentration and the S^{IV} concentration in the rainwater. For situations where the SO_2 concentration in the atmosphere was such that dissolved S^{IV} exceeded 40 μM

$$R_{(SO_4^=)} = k_c[Mn] \qquad (1)$$

and for situations where S^{IV} was less than 40 μM

$$R_{(SO_4^=)} = k_c[Mn] \left[\frac{S^{IV}}{100} \right]^{3/2} \qquad (2)$$

No other catalytic metal was found to directly influence the rate although there may be indirect effects, since the kinetics of the reaction were quite different to those observed with pure reagents in distilled water.

The rate of the uncatalysed reaction of sulphite ions with oxygen was studied for comparison.[53] This process has been studied by many workers who have found that it is extremely susceptible to minute concentrations of both catalysts and inhibitors.[46] Both distilled water and rainwater data in the study undertaken as part of the COST program agreed in indicating a very slow rate which obeyed the overall kinetic scheme

$$R_{(SO_4^=)} = k_{(O_2)} [SO_3^=] \qquad (3)$$

In contrast the rate of bisulphite oxidation by ozone and by hydrogen peroxide were both very fast and had to be studied with a stopped-flow apparatus. The kinetics of the ozone reaction were found to be

$$R_{(SO_4^=)} = k_{(O_3)} [O_3][HSO_3^-][H^+]^{-1/2} \qquad (4)$$

and the kinetics of the hydrogen peroxide reaction were approximately

$$R_{(SO_4^=)} = k_{(H_2O_2)} [H_2O_2][HSO_3^-][H^+] \qquad (5)$$

This rate expression (5), involving peroxide, was the only one which showed a positive response to protons and as such it must be considered as a very important process for creating acidity.

All of the processes investigated above must make some contributions to the production of sulphate and acidity but their relative importance will depend upon circumstances. The figures in Table (1) show that catalytic oxidation can be of importance in situations where the catalyst and SO_2 concentrations are very high. The rates shown in column $R_{(1)}$ were estimated using

Table 1
Table of rainwater data showing pH values, manganese and sulphate
concentrations with calculated SO_2 oxidation rates at 10°C using
equations (1) and (2)

Mn μM	$R_{(1)}$ $\mu M/min$	$R_{(2)}$ $\mu M/min$	SO_4 μM	pH
1.60	1.92	0.186	87.0	4.3
0.78	0.94	0.091	162.0	4.1
0.47	0.56	0.055	46.2	4.2
0.15	0.18	0.017	86.0	4.4
0.07	0.08	0.0081	75.2	3.7
0.06	0.07	0.0070	72.8	4.0
0.04	0.05	0.005	92.3	4.4

equation (1) and the manganese content of the rainwater shown.
In urban areas the manganese content of rain can exceed 5 μM[56]
and the manganese content of fog water can be much higher. It is
very likely therefore that much of the sulphuric acid observed in
urban fogs is produced through catalytic oxidation with manganese
as the main catalyst. In cleaner areas with lower pollutant
concentrations rates shown under $R_{(2)}$ of Table 1 would apply.
These were calculated using Equation 2 and assuming that the
ambient SO_2 concentration was 5 ppbv. They are very small when
considered against the sulphate content of the rainwater and it
seems unlikely that catalytic oxidation could be responsible for
much of the sulphur observed either in rain or as an aerosol
outside urban areas. It is perhaps not surprising therefore that
the catalyst content and the sulphate content of the atmosphere
are unrelated (Figures 6 and 7).
It was pointed out earlier that non-catalytic mechanisms
have been favoured for producing sulphate in rain. This point is
addressed in Table 2 which compares estimated rates of production
of sulphate by reaction of dissolved SO_2 with oxygen, ozone and
hydrogen peroxide. The oxygen rates, estimated from the work of

Table 2
Table showing calculated rates of $SO2$ oxidation by O_2, O_3
and H_2O_2 at 10°C. Initial reactant concentrations are
SO_2 = 5 ppbv, O_3 = 50 ppbv, H_2O_2 = 1 ppbv

pH	O_2 $\mu m/min$	O_3 $\mu m/min$	H_2O_2 $\mu m/min$
4.0	0.0001	0.1	
4.3			13.4
5.0	0.003	2.5	
5.4			28.5
6.0	0.6	67.6	
6.6			73.4

Beilke and his coworkers[57] are very small, below pH 6, the
ozone rates are significant at all pH's above 4 becoming very
large at pH 6 and the peroxy rates are significant at all pH

values. The differences in efficiency of the various rate
processes is perhaps shown more clearly in Figure 10.

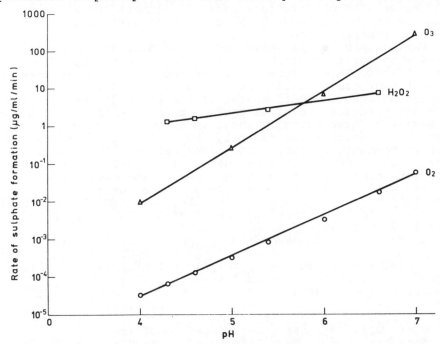

Figure 10   Diagram showing the relative efficiencies of
                   oxidation of SO_2 by O_2, O_3 and H_2O_2
                   under the conditions quoted in Table 2

This is a logarithmic plot of rate vs hydrogen ion concentration.
The oxygen mechanism is slow and sensitive to pH. The ozone
mechanism is very sensitive to pH, becoming dominant in an
alkaline regime. The hydrogen peroxide mechanism is rather
insensitive to pH and is dominant under acid conditions. It is
of course very sensitive to the concentrations of hydrogen
peroxide in the atmosphere, which will depend greatly on
photochemical activity. Thus this mechanism will show a
pronounced seasonal variation by perhaps a factor of 6. An
important conclusion to be drawn from these quantitative
assessments must be that uncatalysed oxidation of SO_2 by ozone or
hydrogen peroxide is probably the dominant droplet phase
mechanism. The attainable rates are quite capable of producing
enough sulphate to account for observed rainwater levels within a
short period of time, especially if the reactant concentrations
are sustained.

2.3   Recent work on problems related to liquid phase oxidation
      processes
      One of the most productive areas of recent work has been the
observation of hydrogen peroxide in the atmosphere. Measurements
made in the gas phase are the subject of some dispute but
measurements made in rain and cloud water by several workers are
now accepted as being real.[58-62] Measurements made on the cloud

water content in Europe by Romer[63] as part of the COST 61a bis program indicate that H_2O_2 is removed by reaction with SO_2. This would agree with observations made of sulphate formation in clouds in the U.S.A. made by Hegg and Hobbs,[64] who argued in favour of a peroxide mechanism.

Various papers also have shown that hydrogen peroxide is formed when normal atmospheric air is bubbled through distilled water.[61,62] This observation confused the gas phase measurements of hydrogen peroxide, but it has also led to the postulation that the main source of the peroxide observed in cloud and rainwater involves aqueous phase chemistry. Chameides and Davis[65] have recently shown in a convincing theoretical paper that hydroperoxide radicals, HO_2, will be formed by gas phase photochemistry inside the cloud medium. These will preferentially be captured by the cloud droplets rather than react with nitric oxide in the gas phase, and having been captured will combine to form hydrogen peroxide inside the droplets. Chameides and Davis[65] also proposed that hydroxyl radicals could be captured by the cloud droplets and react directly with dissolved sulphur dioxide.

Of the oxidation mechanisms proposed only the peroxide mechanism is expected to show a large seasonal variation. Based upon annual measurements of peroxy acetyl nitrate, which is a unique product of photochemical activity, it seems likely that the rate will drop by a factor of 6 in the winter.[18] This still may allow for an appreciable rate all year round, based upon the figures in Table 2, although it is likely that the other mechanisms will dominate under winter conditions.

Some American workers,[66] have proposed that peroxy acetyl nitrate could also act as a good oxidant in atmospheric droplets; it would also give rise to nitrate formation eventually. This is most unlikely however since the compound is not readily taken up by water surfaces. Its measured deposition velocity to water surfaces is less than 0.05 cm sec^{-1}[67] and observations suggest that its concentration in the atmosphere remains stable in the presence of falling rain.[68]

With the realisation that nitric acid is making a larger contribution to atmospheric acidity than before, interest in droplet mechanisms leading to nitrate formation has increased. Based upon the precepts outlined in the introduction to this section, NO_2 is an unlikely candidate to be involved in droplet chemistry. It is not very soluble at atmospheric concentrations since it appears to go into solution as its dimer N_2O_4. There is considerable evidence though that some nitogen and oxygen containing species are very soluble and do give rise to nitrate formation. The growth of nitrate in winter found by Salmon and Atkins et al[51] strongly suggests some non-photochemical and therefore possibly droplet reaction. There are several observations known to the authors of a sudden onset of high nitrate levels in the atmosphere associated with fog formation in the early morning hours.[69] Recent spectroscopic work has indicated the presence of NO_3[70] and HNO_2[71] in polluted air, which are likely to be very soluble and therefore find their way into fog droplets. NO_3, particularly, is a good candidate. It is formed from the reaction of ozone with nitrogen dioxide.

$$O_3 + NO_2 \quad \text{--->} \quad NO_3 + O_2$$

The gas phase removal processes have been discussed above and they are relatively rapid. In the presence of water droplets, however, an alternative fate is capture by the droplets[72] in a similar manner to that proposed for HO_2 radicals.[65] In this event the most probable chemical process that will occur is

$$NO_3 + HSO_3^- \longrightarrow NO_3^- + HSO_3$$

which is very similar to the reaction proposed in the hydroxyl radical oxidation of bisulphite ions[53]

$$OH + HSO_3^- \longrightarrow OH^- + HSO_3$$

The bisulphite radical then reacts with oxygen and is ultimately converted to a bisulphate ion. The energetics of the NO_3 reaction is very similar to the OH reaction and it is likely therefore that the rate of the simple electron transfer process would also be very fast. This process would lead to simultaneous nitrate and sulphate production and could be a process of great significance in the atmosphere. There is already some experimental evidence for the production of nitrate from the O_3 + NO_2 reaction in the presence of a delequescant ammonium sulphate aerosol which was capable of intercepting NO_3 or N_2O_5[73]. There is also evidence that NO_2 itself is not likely to cause much oxidation of SO_2 in solution.[74]

Finally before completing this section it should be pointed out that fog droplet chemistry of the sort that was probably responsible for producing the great urban smoke fogs in Europe is now being investigated vigorously in California. Hoffman and his coworkers at the California Institute of Technology are finding evidence for catalytic oxidation of SO_2 in droplets under these circumstances.[75] Also Novakov and his coworkers at the Lawrence Berkely Laboratory are investigating the potential of wet soot particles for oxidising SO_2[76].

3.   OXIDATION ON THE SURFACE OF ATMOSPHERIC PARTICULATE
This is a subject where there is very little hard information to date on which to assess its applicability under atmospheric conditions. At various times there have been people who have advocated this mechanism, the most recent case being Novakov[77] and his coworkers who claim that oxidation on soot particles could account for the sulphate aerosol observed in Los Angeles. In view of what is known about the efficiency of the gas phase processes involving hydroxyl radicals and Novakov's recent views on the importance of a liquid film on the particles[76], this would seem to be very unlikely though. Liberti et al[78], who have spent much time investigating the surface reactions of sulphur dioxide, concluded in a recent review that 'Conversion to sulphate at room temperature does not take place to a considerable extent on atmospheric dusts unless aerosols due to industrial emissions of a specific nature affect their composition'.

It is possible then that oxidation of this type may occur in special situations, such as in the plume of a chimney very close to the source, but it is also very unlikely that this reaction is important in the atmosphere at large.

4.   RESEARCH NEEDS

This review has indicated that much information is known about the formation of acid substances in the atmosphere. A lot remains unknown however, particularly concerning processes that take place in the liquid phase. The most efficient of these involve photochemistry which will undoubtedly show a large seasonal variation. In order to account for the observations of sulphate and nitrate in winter, there are almost certainly efficient thermal processes and these need to be correctly identified.

A novel mechanism which may supplement the O_3/H_2O_2 oxidation of SO_2 in solution, involves chlorine nitrate. As mentioned above it is formed from the oxidation of HCl in the gas phase by hydroxyl radicals. It hydrolyses in water to produce a very powerful oxidising agent, hypochlorous acid. The potential influence of this process needs to be quantified.

It is possible that formaldehyde can inhibit SO_2 oxidation in droplets by producing an aldehyde-bisulphite complex. Formaldehyde is produced by photochemical oxidation of hydrocarbons in the atmosphere and thus the influence of this inhibition process should be much less in winter. Nothing is known about this at present though.

The effects that emissions of pollutants from motor vehicles can have on acid deposition needs to be investigated in detail. Motor vehicles are a direct source of oxides of nitrogen that can be converted into nitric acid of course. They are also a source of reactive hydrocarbons and aldehydes which can influence the rate of production of substances in the atmosphere.

New methods of oxidation involving free radicals and cloud droplets have been proposed that need quantification. Some chemical processes which have been known for many years have still to be evaluated under conditions that apply in the atmosphere. This is particularly true in the case of carbonaceous aerosols that may present an efficient thermal route to the production of sulphuric acid from SO_2.

The whole field of NO_x chemistry in the liquid phase is largely unknown but there are strong pointers that it could be of considerable significance. The critical role of NO_x in controlling the homogeneous oxidation rates has been known for several years. Plumes present a special case where high NO levels may suppress oxidation, also high SO_2 may cause the hydroxyl radical concentration to be suppressed by forming sulphuric acid in a radical addition process.

The potential effects of acid deposition from fossil fuel combustion emissions can only be assessed properly against a base line of the natural deposition of acidity. It is therefore desirable to define the main vectors for sulphur and nitrogen in the unperturbed atmosphere and to determine their rate of conversion to acid substances.

A list of research needs in this field is appended below.
1.   Seasonal variations in rates of homogeneous and liquid phase processes.
2.   Scavenging of free radicals by cloud droplets - origin of oxidants - H_2O_2 etc - in rain water.
3.   Mechanism of OH + SO_2 reaction and its effect on HO_x budget in plumes at far field.

4. NO_3 chemistry and heterogeneous/aqueous - phase reactions of NO_x.
5.   Natural component of acid rain. Contributions of CS_2 and DMS to SO_2 budget; budge of CH_3SO_4H from DMS.
6.   Role of other pollutants in affecting local oxidation rates e.g. OH dependence on hydrocarbons.
7.   Carbonaceous aerosols - their role in aqueous phase oxidation.
8.   The role of chlorine chemistry in the formation of atmospheric acidity.
9.   Potential inhibition of liquid phase oxidation of SO_2 by formaldehyde.

Acknowledgement

This work was supported by the UK Department of the Environment.

References
(1)   C.E. Junge and T.G. Ryan, 'Study of the SO_2 oxidation in solution and its role in atmospheric chemistry', Quart. J. Roy. Met. Soc. 84 46-55 (1958).
(2)   P.A. Leighton, 'Photochemistry of Air Pollution', Acad. Press. N.Y. (1961)
(3)   R.A. Cox and R.G. Derwent, 'Gas Phase Chemistry of the Minor Constituents of the Troposphere' in Royal Society of Chemistry SPR 'Gas Kinetics and Energy Transfer', 4 189-235 (1981).
(4)   S.A. Penkett, 'Oxidation of SO_2 and other Atmospheric Gases by Ozone in Aqueous Solution, Nature 240 105 (1972).
(5)   D.H.F. Atkins, R.A. Cox and A.E.J. Eggleton, 'Photochemical Ozone and Sulphuric Acid formation in the atmosphere over Southern England', Nature 235 372-375 (1972).
(6)   A.E.J. Eggleton and R.A. Cox, 'Homogeneous Oxidation of Sulphur Compounds in the Atmosphere', Atm. Environ. 12 227-230 (1978).
(7)   W.E. Clark, D.A. Landis and A.B. Harker, 'Measurements of the Photochemical Production of Aerosols in Ambient Air near a freeway for a range of SO_2 concentrations', Atmos. Environ. 10 637-644 (1976).
(8)   J.G. Calvert, FuSu, J.W. Bottenheim and O.P. Strausz, 'Mechanisms of the homogeneous oxidation of Sulphur Dioxide in the Troposphere'. Atmos. Environ. 12 (1978).
(9)   D.L. Baulch, R.A. Cox, P.J. Crutzen, R.F. Hampson, J.A. Kerr, J. Troe and R.T. Watson, CODATA Task Group on Chemical Kinetics, 'Evaluated Kinetic and photochemical Data for Atmospheric Chemistry: J. Phys. Chem. Ref. Data, Supplement I, 11 327-496 (1982).
(10)   R.A. Cox and S.A. Penkett, 'Oxidation of SO_2 by oxidants formed in the Ozone-Olefin reaction', Nature 230 321 (1971).
(11)   R.A. Cox and S.A. Penkett, 'Aerosol formation from Sulphur Dioxide in the presence of Ozone and Olefinic Hydrocarbons', J. Chem. Soc. Faraday Trans I 68 1735-1753 (1972).

(12)   T.N. Rao, S.S. Collier and J.G. Calvert, 'The Quenching
         Reactions of the First Excited Singlet and Triplet States
         of Sulfur Dioxide with Oxygen and Carbon Dioxide', J. Am.
         Chem. Soc. 91 1616 (1969).
(13)   L. Newman, 'Atmospheric Oxidation of Sulphur Dioxide: A
         Review as viewed from power plant and Smelter Plume
         Studies', Atmos. Environ. 15 2231-2239 (1981).
(14)   R.G. Derwent and O. Hov, 'Computer Modelling Studies of
         Photochemical Air Pollution formation in North West
         Europe', AERE Report R-9434 (1979)
(15)   H.W. Sidebottom, C.C. Badcock, G.E. Jackson, J.G.
         Calvert, G.W. Reinhardt and E.K. Damon, 'Photo-oxidation
         of Sulphur Dioxide', Environ. Sci. Technol. 6 72 (1973).
(16)   P.H. Wine, W.L. Chameides and A.R. Ravishankara,
         'Potential Role of CS_2 Photo-oxidation in Tropospheric
         Sulphur Chemistry', Geophys. Res. Lett. 8 543-546 (1981).
(17)   T. Okita and S. Ohita, in 'Nitrogenous Air Pollutants –
         Chemical and Biological Implications', ed. D. Grosjean,
         Ann. Arbour. Science, Michigan 1979, pp189-306.
(18)   S.A. Penkett, F.J. Sandalls and B.M.R. Jones, 'PAN
         Measurements in England – Analytical Methods and
         Results', VDI-Berichte No 270, 1977.
(19)   E. Robinson and R.C. Robbins: Sources, Abundance and Fate
         of Atmospheric Pollutants – Supplement', American
         Petroleum Institute Publication No 4015, April 1971.
(20)   D.H. Ehhalt and J.W. Drummond, 'Proceedings of NATO
         Advanced Study Institute on Chemistry of the Unpolluted
         and Polluted Troposphere, Corfu September-October 1981.,
         D. Reidel Publ. Co., Dordrecht (1982)
(21)   D. Kley, J.W. Drummond, M. McFarland and S.C Lui,
         'Tropospheric Profiles of NO_x', J. Geophys. Res. 86 3153-
         3161 (1981).
(22)   V. Platt, D. Perner, G.W. Harris, A.M. Winer and J.N.
         Pitts, 'Observation of Nitrous Acid in an Urban
         Atmosphere by Differential Optical Absorption', Nature
         285 312-314 (1980).
(23)   R.A. Cox, R.G. Derwent and A.J.L. Hutton, 'Significance
         of Peroxynitric Acid in Atmospheric Chemistry of Nitrogen
         Oxides, Nature 270 328-329 (1977).
(24)   R.A. Cox and M. Roffey, 'Thermal decomposition of
         Peroxyacetylnitrate in the presence of nitric oxide',
         Environ. Sci. Technol. 11 900-906 (1977).
(25)   V. Schirath and V. Wipprecht, 'Reactions of peroxyacyl
         Radicals', in Proceedings of First European Symposium –
         Physico-chemical behaviour of Atmospheric Pollutants,
         Ispra 16-18 October 1979, Ed. B. Versino and H. Ott, CEC
         publications EUR 6621, 1980, pp.157-166.
(26)   P.J. Crutzen, 'NO_x in the Stratosphere and Troposphere',
         Ann. Rev. Earth and Planet Sci. 76 43-70 (1979).
(27)   F.S. Rowland and M.J. Molina, 'Chlorofluoromethanes in
         the Environment', Rev. Geophys. Space Physic 13 1-
         (1975).
(28)   M. Schmeisser and K. Brandle, 'Halogennitrate und ihre
         reactionen', Angew. Chem. 73 388-393 (1961).
(29)   J.H. Coste and G.B. Courtier, 'Sulphuric Acid as a
         disperse phase in town air', Trans. Far. Soc. 32 1198
         (1936).

(30)   M. Hoffman, 'California Instute of Technology, personal
        communication (1982).
(31)   H.F. Johnstone, 'Metallic ions as catalysts for the
        removal of sulphur dioxide from boiler furness gas', Ind.
        and Eng. Chem. $\underline{23}$ 559 (1931).
(32)   R.C. Hoather and C.F. Goodeve, 'The oxidation of
        sulphurous acid III Catalysis by manganous sulphate',
        Trans. Far. Soc. $\underline{30}$ 67 (1934).
(33)   H. Bassett and W.A. parker, 'The oxidation of sulphurous
        acid', J. Chem. Soc. 1540 (1951).
(34)   H.F. Johnstone and D.F. Coughanower, 'Absorption of
        sulphur dioxide from air', Ind. Eng. Chem. $\underline{50}$ 1169
        (1958).
(35)   H.F. Johnstone and A.J. Moll, 'Formation of sulphuric
        acid in Fogs', Ind. Eng. Chem. $\underline{52}$ 861 (1960).
(36)   D.F. Coughanower and F.E. Krause, 'The reaction of SO_2
        and O_2 in aqueous solutions of Mn SO_4, I and E.C.
        Fundementals $\underline{4}$ 61 (1965).
(37)   J.M. Bracewell and D. Gall, 'Catalytic oxidation of
        sulphur dioxide in solutions at concentrations occurring
        in fog droplets', Proc. Symp. on Phisico-Chemical
        Transformations of sulphur compounds in the Atmosphere
        and Formation of Acid Smogs, Mainz, Germany (1967).
(38)   M.J. Matteson, W. Stober and H. Luther, 'Kinetics of the
        oxidation of sulphur dioxide by aerosols of manganese
        sulphate', I and E.C. Fundementals $\underline{8}$ 677 (1969).
(39)   R, Brodzinsky, S.G. Chang, S.S. Markowitz and T. Novakov,
        'Kinetic and mechanism for the catalytic oxidation of
        sulphur dioxide in carbon in aqueous suspensions', J.
        Phys. Chem. $\underline{84}$ 3354 (1980).
(40)   P. Brimblecombe and D.J. Spedding, 'The catalytic
        oxidation of micromolar aqueous sulphur dioxide. I.
        Oxidation of dilute solutions containing iron (III)',
        Atmos. Environ. $\underline{8}$ 937 (1974).
(41)   J. Freiberg, 'The mechanism of iron catalysed oxidation
        of SO_2 in oxygenated solutions', Atmos. Environ. $\underline{9}$ 661
        (1975).
(42)   L.A. Barrie, and H.W. Georgii, 'An experimental
        investigation of the absorption of sulphur dioxide by
        water drops containing heavy metal ions', Atmos. Environ.
        $\underline{10}$ 743 (1976).
(43)   W.I. Scott, and P.V. Hobbs, 'The formation of sulphate in
        water droplets', J. Atmos. Sci. $\underline{24}$ 54 (1967).
(44)   A.P. Van den Heuval and B.J. Mason, 'The formation of
        ammonium sulphate in water droplets exposed to gaseous
        sulphur dioxide and ammonia', Quart. J. Roy. Met. Soc. $\underline{89}$
        271 (1963).
(45)   H.A.C. McKay, 'The atmospheric oxidation of sulphur
        dioxide in water droplets in the presence of ammonia',
        Atmos. Environ. $\underline{5}$ 7 (1971).
(46)   E.C. Fuller and R.H Crist, 'The rate of oxidation of
        sulphite ions by oxygen', Am. Chem. Soc. J. $\underline{63}$ 444
        (1941).
(47)   S.A. Penkett and J.A. Garland, 'Oxidation of sulphur
        dioxide in artificial fogs by ozone', Tellus $\underline{26}$ 284
        (1974).

(48)   R.E. Erikson, L.M. Yates, R.L. Clark and D. McEwen, 'The
         retention of sulphur dioxide with ozone in water and its
         possible atmospheric significance'.
(49)   T.V. Larson and H. Harrison, 'Acidic sulphate aerosols:
         Formation from heterogeneous oxidation of O_3 in clouds',
         Atmos. Environ. 11 1133 (1977).
(50)   D.A. Hegg and P.V. Hobbs, 'Oxidation of sulphur dioxide
         in aqueous systems with particular reference to the
         atmosphere', Atmos. Environ. 12 241 (1978).
(51)   L. Salmon, D.H.F. Atkins, E.M.R. Fisher, C. Healy and
         D.V. Law, 'Retrospective trend analysis of the content of
         U.K. air particulate material 1957–1974', The Science of
         the Total Environment 9 161 (1978).
(52)   P. Brimblecombe and P.H. Stedman, 'Historical evidence
         for a dramatic increase in the nitrate component of acid
         rain', Nature 298 460 (1982).
(53)   S.A. Penkett, B.M.R. Jones, K.A. Brice and A.E.J.
         Eggleton, 'The importance of atmospheric ozone and
         hydrogen peroxide in oxidising sulphur dioxide in cloud
         and rainwater', Atmos. Environ. 13 123 (1979).
(54)   S. Beilke and G. Gravenhorst, 'Heterogeneous SO"2'
         oxidation in the droplet phase', Atmos. Environ. 12 231
         (1978).
(55)   S.A. Penkett, B.M.R. Jones and A.E.J. Eggleton, 'A study
         of SO_2 oxidation in stored rainwater samples', Atmos.
         Environ. 13 139 (1979).
(56)   A. Meurrans and Y. Lenelle, 'Sulphides and sulphates in
         rain: role of rain in the elimination of atmospheric
         SO_2', Report from Institut d'Hygiene et d'Epidemiologie,
         Brussels (1976).
(57)   S. Beilke, D. Lamb and J. Muller, 'On the uncatalysed
         oxidation of atmospheric SO_2 by oxygen in aqueous
         systems', Atmos. Environ. 9 1083 (1976).
(58)   G.L. Kok, 'Measurement of hydrogen peroxide in
         rainwater', Atmos. Environ. 14 653 (1980).
(59)   G.L. Kok, K.R. Darnoll, A.M. Winer, J.N. Pitts and B.W.
         Gay, 'Ambient air measurements of hydrogen peroxide in
         the California South Coast Air Basin', Environ. Sci.
         Tech. 12 1077 (1978).
(60)   T.J. Kelly, D.H. Stedman and G.L. Kok, 'Measurements of
         H_2O_2 and HNO_3 in rural air', Geophys. Res. Lett. 6 375
         (1979).
(61)   B.A. Heikes, A.L. Lazrus, G.L. Kok, S.M. Kanen, B.W.
         Gandrud, S.N. Aitling and P.D. Sperry, 'Guideline for
         aqueous phase hydrogen peroxide synthesis in the
         troposphere' J. Geophys. Res. 87 3045 (1982).
(62)   R.G. Zika and E.S. Saltzman, 'Interaction of ozone and
         hydrogen peroxide in water: implications for analysis of
         H_2O_2 in air', Geophys. Res. Letts. 9 231 (1982).
(63)   F.R. Romer, COST symposium on Acid Deposition, Berlin
         (1982).
(64)   D.A. Hegg and P.V. Hobbs, 'Cloud chemistry and the
         production of sulphate in clouds', Atmos. Environ. 15
         1596 (1981).
(65)   W.L. Chameides and D.D. Davis, 'The free radical
         chemistry of cloud droplets and its impact upon the
         composition of rain', J. Geophys. Res. 87 4863 (1982).

(66)   Organic nitrogen compounds can affect rainfall acidity.
         News article in J.A.P.C.A. <u>32</u> 1075 (1982).
(67)   J.A. Garland and S.A. Penkett, 'Absorption of peroxy
         acetyl nitrate and ozone by natural surfaces', Atmos.
         Environ. <u>10</u> 1127 (1976).
(68)   K.A. Brice and S.A. Penkett, unpublished observations.
(69)   D.H.F. Atkins, unpublished observations.
(70)   U. Platt, D. Perner, A.W. Winer, G.W. Harris and J.N.
         Pitts, 'Detection of NO_3 in the polluted troposphere by
         differential optical absorption', Geophys. Res. Lett. <u>7</u>
         89 (1980).
(71)   U. Platt and D. Perner, 'Direct measurement of
         atmospheric CH_2O, HNO_2, O_3, NO_2 and SO_2 by differential
         optical absorption in the near U.V.', J. Geophys. Res. <u>85</u>
         7453 (1980).
(72)   U. Platt, D. Perner, J. Schroder, C. Kessler and
         Toennissen, 'The diurnal variation of NO_3', J. Geophys.
         Res. <u>86</u> 11965 (1981).
(73)   R.A. Cox, 'Particle formation from homogeneous reactions
         of sulphur dioxide and nitrogen dioxide', Tellus <u>26</u> 235
         (1974).
(74)   L. Robbin Martin, D.E. Damschko and H.S. Judeikis, 'The
         reactions of nitrogen oxides with SO_2 in aqueous
         aerosols,', Atmos. Environ. <u>15</u> 191 (1981).
(75)   J.M. Waldman, J.W. Munger, D.J. Jacob, R.C. Flagan, J.J.
         Morgan and M.R. Hoffman, ' Chemical composition of acid
         fog', Science, <u>218</u> 677 (1982).
(76)   W.H. Benner, R. Brodzinsky and T. Novakov, 'Oxidation of
         SO_2 in droplets which contain soot particles', Atmos.
         Environ. <u>16</u> 1333 (1982).
(77)   T. Novakov, S.G. Chang and A.B. Harker, 'Sulphates as
         pollution particulates: catalytic formation on carbon
         (soot) particles', Science <u>186</u> 259 (1974).
(78)   A. Liberti, D. Brocco and M. Possanzini, 'Absorption and
         conversion of sulphur dioxide on particles', Atmos.
         Environ. <u>12</u> 255 (1978).

BACKGROUND PRECIPITATION ACIDITY

R.J. DELMAS and G. GRAVENHORST
Laboratoire de Glaciologie et Géophysique de l'Environnement

Summary

Existing data concerning the precipitation chemistry of unpolluted
areas are reviewed in order to better understand background acidity.
The knowledge of this parameter is relevant for assessing the extent
of anthropogenic acid rain phenomenon to the remote regions of the
world. We describe first the different techniques generally used for
acidity measurements. It is shown that rain and snowmelt water samples
collected in background conditions should be analyzed with particular
precautions. Then several examples of the pH (or acidity) values of
precipitation obtained in maritime, continental and polar unpolluted
regions are discussed. We are led to the conclusion that the pH value
of 5.6 (pure water in equilibrium with atmospheric CO_2) cannot be
taken as a natural reference pH of rainwater because atmospheric acid
and basic trace constituents modify significantly this figure, depending on the regional budget of their respective emission rates into
and removal rates from the atmosphere. Gas-derived secondary sulfur
and nitrogen constituents of aerosols are the most important trace
substances that acidify background precipitation. In marine air conditions, sea-salt particles, and over continents, terrestrial dust and
NH_3 may neutralize significantly these natural acids, at least at
lower altitudes. In central polar regions the importance of all major
aerosol sources is considerably reduced so that ultra low background
concentration levels of acidity are generally observed in polar snow.
Finally acid precipitation formation is largely discussed in the light
of our knowledge of major atmospheric trace substances (aerosol and
gases).

1. INTRODUCTION

Vast areas of the Northern Hemisphere, including Central Europe,
Scandinavia, and North-East of U.S.A. and of Canada are affected by acid
pollution. There is a broad concern that acidity of rain can have significant environmental impacts, in particular on the aciditication of lakes
and on plant-soil systems. At first glance, it may appear surprising that
certain affected areas are not characterized by particularly high levels
of industrialization. Investigations carried out over the last decade have
shown that large amounts of anthropogenic compounds are emitted into air
masses passing over industrial zones and that they can subsequently be
transported several thousand kilometers away. The networks used in monitoring precipitation chemistry have therefore been extended to the so-called
"background" regions of the globe, regions where the anthropogenic influence
should be reduced to insignificant levels. Very little work had been done in
the past on precipitation chemistry in these areas, and it is now becoming
increasingly evident that the composition of rainwater and snow is of great
value in evaluating natural and anthropogenic contributions to atmospheric

cycles of trace elements. This paper reviews results which are presently available on the acidity of precipitation in background regions and tries to pick out reasons for a possible natural acidity in rainfall and snow.

Note that one can make a difference between the concepts of"background acidity" and "reference level acidity". The acidity reference level at a given site relates to the level without any influence of human activity. Continental background refers to stations where contributions from marine and anthropogenic sources are minimal. Correspondingly, marine background refers to stations where contributions from continental and anthropogenic sources are minimal. It should be emphasized that these two concepts are combined in polar regions due to the possibility of determining pre-industrial levels and due to the absence of neighbouring primary aerosol sources. Particularly in Central Antarctica-marine, soil and anthropogenic sources are absent. Glaciochemistry, represents therefore an elegant and original method of reconstituting background acidities of precipitation over several tens of thousands of years, since the acidity of past snow is kept unchanged within ice sheets or glaciers usually located in remote areas.

Pure water has a pH of 7. Rain-water, is however not made up of pure water. The droplets formed from atmospheric water vapor are loaded with substances coming from their surrounding environment, either during cloud drop formation and in-cloud scavenging (rain-out processes) or as the droplets fall towards the ground (wash-out process). Moreover, cloud and rain-water absorbs atmospheric trace gases (e.g. CO_2, SO_2, NH_3, etc.) which alter its chemical composition. In particular omnipresent carbon dioxide dissolves to form carbonic acid, which in turn is capable of lowering the pH of rainwater from 7 to about 5.6 (the equilibrium pH of water with an atmosphere containing 330 ppvm of CO_2). In reality this value of 5.6 is rarely observed as it is significantly modified by other soluble compounds, either acidic or basic.

It is known that atmospheric rain-water chemistry is largely dominated by the reaction of sulfur and nitrogen compounds (Junge, 1963 a). These compounds (e.g. SO_2, NO_x, $(CH_3)_2S$, COS) are especially capable of forming acids via homogeneous and/or heterogeneous reactions. The only important alkaline trace gas seems to be NH_3. In the atmospheric budget of these sulfur and nitrogen gases a considerable role is played nowadays by anthropogenic emissions (Granat et al., 1976 ; Böttger et al, 1978 ; Janssen-Schmidt et al., 1981). If one assumes only natural emission processes at the earth surface the production rates of total gaseous sulfur and nitrogen compounds would be largely reduced. Even, these natural fluxes into the atmosphere would, however, decrease in general the rain-water pH below a value of 5.6. Considering only the atmosphere cycle a pH range of 4.5 to 5.6 was deduced for average natural precipitation (Charlson and Rodhe, 1982). Over remote continental areas, however, alkaline airborne components such as calcareous dust and biogenic ammonia can significantly increase the pH of rain-water to values even greater than 7. Since the atmospheric circulation can transport trace constituents around over large distances (several thousands of km) the background free acidity in rain-water will be the result of multiple natural meteorological and physico-chemical phenomena in the atmosphere as emphasized by Sequeira (1982). The observed background pH value will therefore vary considerably from place to place and from time to time (Figure 1).

In the following it will be first attempted to briefly describe the techniques generally used to determine acidity of precipitation samples. Then selected precipitation pH data obtained in various unpolluted regions of the world will be presented.

From that limited survey of rainchemistry, some general aspects of background pH distribution will be discussed.

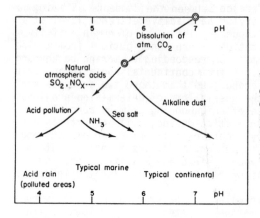

Figure 1

The pH range predicted for background precipitation is wide, depending on the relative concentrations of the various atmospheric trace constituents that can be incorporated in water droplets or snowflakes

2. METHODS FOR MEASURING PRECIPITATION ACIDITY

2.1. Rainwater

Acidity measurements are generally made by pH determination or by titration. Each method, however, determines a different parameter. pH measurement gives the concentration (in mole l^{-1}) of H^+ ions (pH = $- \log H^+$) ; that means the effective concentration of free protons in the solution under investigation. Titrations (the titrimetric procedures are multiple) determine the total acid neutralizing capacity ; that means the total proton availability in the system until the endpoint pH.

A variety of components such as inorganic or organic, weak or strong acids, or metal oxides (e.g. those of Al or Fe) may contribute to the free acidity or fain. pH method does not give any indication about the acid-base system responsible for the observed free acidity whereas titrimetric methods allow the separation of the respective contributions of the weak and strong acids (Askne and Brosset, 1972). Further, if a refined titration method is used (Gran, 1952), we can even obtain the amount and the strength of each of the weak acids involved.

Acidity may also be estimated indirectly by determining the concentrations of all principal soluble anions and cations in the solution (Granat, 1972). In this case, the sum (concentrations expressed in equivalents l^{-1} of the cations minus the sum of the anions is taken as the concentration of free H^+ ions).

A coulometric method was used by Liberti et al. (1972) to measure the non-volatile acidity of rainwater. An elegant radio-chemical method was developed by Klockow et al. (1978) to determine strong acids. In that method, the acid rain sample is allowed to react with radioactive $NaCl^*$. The amount of volatile HCl^* formed corresponds exactly to that of strong acids present in the sample.

It is important to note that ionic concentrations in precipitation water are often very low in background regions, which may create problems in pH measurement due to the high electrical resistivity of the solution. Furthermore, certain precautions must be taken in sample-handling to avoid contamination. An extreme case is that of polar snow and ice, where accurate analysis is much more difficult. But, important analytical errors may be made even in determining the pH of rainwater from temperate regions as is demanstrated by the intercomparison of results obtained in several laboratories under carefully controlled conditions (Table 1), showing differences of up to one pH unit for the same sample.

N° of laboratories	Known value	Mean of values reported	Standard deviation (σ)
---------------------------- pH -------------------			
18	6.07	5.56	0.41
18	6.15	5.53	0.76
17	6.18	5.45	0.74
17	6.20	5.54	0.52
--------------------- $H^+ \mu Eq. \ 1^{-1}$ ------------			
21	37.6	63.5	21.2
18	52.5	25.3	63.7
21	59.9	99.0	27.5

Table 1 : Results of an interlaboratory comparison of rainwater sample acidity measurements (pH determinations or titration). (Tyree, 1981)

To this, it is necessary to add errors related to rainwater sampling, transport and conservation conditions. Furthermore, some acid deposition networks collect only rainfall, while others collect both wet and dry deposits (open rain gauges). In interpreting published data, one must therefore be extremely critical and carefully examine sampling, transport, storage and analysis conditions.
Finally, it must be kept in mind that in certain cases, the insoluble fraction may influence the acid-base equilibria occuring in precipitation samples and also that biological activity can transform rainwater constituents to produce acids.

2.2. Polar precipitation

Little is yet known concerning the acidity of polar precipitation. Special methods have recently been developed either to measure it directly in the ice or in melted samples.

2.2.1. Millar (1981) proposed a remote sensing method based on radar soundings of the ice sheet to reveal layers of "acidic" ice, in particular those containing volcanic fall-out. This method, which has only been calibrated approximately up to now, would be especially valuable in detecting occurences of explosive volcanic activity, recorded over the entire thickness of the polar ice caps. However, it is not certain that acidity is the only cause of observed radar echoes.

2.2.2. The technique developed by Hammer (1980) is based on the electrical conductivity of solid ice, measured by means of two electrodes which are slid along the surface of ice cores. The acidity of the ice is proportional to its conductivity. This method was calibrated and successfully applied to several hundred meters of core samples from Greenland and Antarctica in order to locate sulfuric acid fall-out related to explosive volcanism.

2.2.3. The free acidity of ice (or snow) may also be determined by analyzing its meltwater. It must however be remembered that the melted water can absorb atmospheric CO_2 and be acidified readily (see introduction). A correction for this effect is necessary to obtain the "true" acidity of ice. However this correction is difficult to estimate and may be a source of error in the range of H^+ concentrations generally involved (1-10 μEq. l^{-1}). A highly accurate titration method ($\pm$ 0.2 μEq. l^{-1}) that overcomes this difficulty has recently been proposed (Legrand et al., 1982). It has also been shown that there is a high risk of contamination (especially from alkaline dust) during sample handling and measurement. This explains why there are still few reliable values of the acidity of polar ice to be found in published reports.

3. RAINFALL ACIDITY IN BACKGROUND REGIONS

3.1. Rainfall acidity on Amsterdam Island

This island (37°50'S, 77°34'E), in the south of the Indian Ocean, can be considered to be representative of purely maritime conditions. The station is part of the Global Precipitation Chemistry Network. Rainfall is collected at 2 meters above the ground at a site located 2 km from the coast and at an elevation of 200 meters. A report on one year of precipitation data appeared recently (Galloway et al. 1982)(Figure 2).

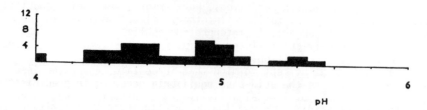

Figure 2 : Histogram of rainwater pH at Amsterdam Island
(from Galloway et al., 1982)

The volume-weighted mean pH was 4.93. One striking feature is that all the samples (36 collected) had a pH less than 5.6 (the pH of pure water in equilibrium with atmospheric CO_2, see Fig. 1) and that the majority of values were within the pH limits calculated by Charlson and Rodhe (4.5-5.6). Two samples had pH values less than 4 (H^+ > 100 μEq. l^{-1}). The approximately one year of precipitation record showed no temporal trend of pH value.

Chemical analysis showed that precipitation on Amsterdam Island contained sea-salt (volume weighted mean Na^+ concentration : 0.18 meq. l^{-1}) and an acidic component. Sulfuric and nitric acids alone did not provide a full explanation for the acid fraction. An additional acid, most likely organic, was suggested to account for the observed pH.

3.2. Rainfall acidity on the island of Hawaii

Since 1974 the pH value of numerous rainfall samples has been determined on the island of Hawaii (Miller, 1980 ; Miller and Yoshinaga, 1981 ; Harding and Miller, 1982 ; Sequeira, 1982). The average pH of 5.2 for rainwater sampled at sea level is similar to those obtained for Amsterdam Island (§ 3.1.) and for Samoa (§ 3.3.). The pH values decrease, however, when rain-water is sampled at higher altitudes and reach an average value of 4.3 at 2500 m a.s.l. (Figure 3).

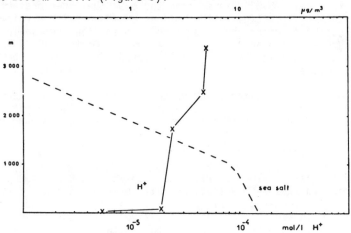

Figure 3 : Rainwater acidity at Hawaii in the function of elevation (m). The stippled line represents the typical sea-salt atmospheric content ($\mu g/m^3$) over the ocean.

This decrease of rainwater pH over the ocean with increasing altitude can be compared with the general decrease of the sea-salt concentration in the aerosol found in many field studies (Blanchard and Woodcock, 1980). This suggests that the degree of neutralization of acidic components by sea-salt (which contains alkaline compounds) decreases with increasing altitude. The volcanic activity on the island was a candidate for rainwater acidification (Harding and Miller, 1982). However it was shown that measurements on Kauai Island (500 km North of Hawai, free of volcanic gas emissions) gave similar pH values. Miller (1980) and Miller and Yoshinaga (1981) examined the possibility of a long range transport of pollutants from the North American continent.

3.3. Rainfall acidity at other marine sites

Rainfall is also acidic on the Bermudas, with an observed-annual average pH value of 4.74, mostly due to H_2SO_4 (Jickills et al., 1982). On

the South coast of Florida, a generally unpolluted area, average pH value
of rainwater was 5.5 (Brezonik et al., 1980) whereas in Northern Florida
north air flow rains exhibited polluted and continental characteristics
(average pH, 4.4) and south air flow rains, originating in the Gulf of
Mexico, had an average pH of 5.3 (Tanaka et al., 1980). Emissions from the
Continental United States seem to influence precipitation acidity on the
Bermudas, situated $\sim$ 1000 km from the American coast. Such sites can no
longer be considered to be clean sites. Generally speaking, it is now difficult to find stations in the Northern Hemisphere which are not affected by
sulfur and nitrate pollution. Even in easter by air flow from the open
North Atlantic, the pH of rain on Bermudas was reported to be $\sim$ 4.9. This
low value of pH supports the relatively large concentration of excess sulfate and the acidity in aerosol samples collected over the North Atlantic
(Gravenhorst, 1975 a, b ; Gravenhorst, 1978).

In the Pacific, on the Island of Samoa, Pszenny et al. (1982), measured
pH values (mean value 5.53) higher than those at the three preceding marine
sites. The authors investigated the influence of sea-salt on rainwater pH
and found that it is minor and evident only in the few most saline samples.

The sources of the acids were not discussed. On the same island,
however, at Cap Matatula, Sequeira (1981) found also pH values between 5
and 6 and suggested that the pH may be controlled by· an unidentified weak
acid.

At Cape Grim , in Tasmania, pH values determined in 1977 were between
5.6 and 6.8 (Sequeira, 1981). These relatively high values appeared to be
a result of an influx of continental material to this background region.

3.4. Rainwater acidity in continental regions

In India, the pH values of rainwaterr between 1975 and 1980, given as
annual averages of the BAPMON stations fell between 5.5 and 8.7 (Government
of India, 1982). Even at the island stations of Port Blair and Minicoy on
the east of the Indian sub-continent, relatively high pH values (6.4 and
6.6 - 6.9 respectively) are measured in this network. On the west coast
(at Thumba, Bombay and Kalliampur) Sequeira (1981) reported pH values of
5.8 to 6.6 for the year 1975.

It seems that the entire area is under the influence of eolian dust
blown up from the arid soils in Central Asia and on the Indian sub-continent (Gravenhorst et al., 1980 ; Georgii, 1982). It is difficult, however,
to get annual pH-averages in the range 7-8 solely by the incorporation of
soluble carbonates into the cloud and rain droplets aloft. The concentration of calcium in rain at New Dehli (Khemani and Rama Murty, 1968) did
not seem to be higher than the ones reported in the United States (for
example at Rio Grande, Texas, Fanning and Lyles, 1964). Other alkaline
components, such as gaseous ammonia evolved from cattle excrements or from
soil organic matter (Böttger et al., 1978) or the absence of acidic sulfur
and nitrogen compounds, could maintain the Indian rainwater pH so high.
The net biogenic productivity in the Indian Ocean is relatively small
(Bonsang, 1982) so that the oceanic contribution to continental aerosol
acidity could be minimal in India. All the speculations for explaining these
high pH values should, however, be postponed until the pH values in Indian
rainwater have been confirmed by a thorough check.

In North America too, relatively high pH are generally reported for
precipitation collected in continental regions. At Alamosa, situated in the
High Plains of the Rocky Mountains (South Colorado), the average pH between
1973 and 1975 was 6.5 (with extremes of 4.6 and 7.9) (Sequeira, 1982). On
the other hand in Northern Colorado,Grant and Lewis (1982) documented a mean

pH value of 4.8 for bulk precipitation samples collected in the Rocky Mountains 40 km North-west of Denver. This apparent discrepancy between the pH data of North and South Colorado precipitation could be linked to different soil conditions as well as to different sampling or analytical procedures. The very rough geographical pH distribution given by Gravenhorst et al. (1980) indicates that in this area a transition occurs from high pH values, observed in the arid and semi-arid regions of the South-west of the United States, to lower pH values, obtained in areas where the ground is covered by vegetation (grass or forests). Furthermore, in the Northern part of the Rocky Mountains, rocks have a non calcareous nature.

In Canada, the Eastern Provinces receive rainwater with pH usually rnaging between 4.2 and 5.6 whereas in Mid-continental Canada the pH values (between 5 and 6.9) indicate more alkaline influences (Barrie, 1981). In the taïga of Alaska, precipitation seems to be acidic. A volume weighted mean pH value of 4.96 is reported for 16 rain events at Poker Flat (67°N, 147°W) since December 1979 (Galloway et al., 1982).

In the large Amazonian basin in South America, rainwater pH ranged between 4.7 and 5.7 (Stallard and Edmond, 1981). In the tropical rain forest of Venezuela, at the Northern edge of the Amazonian basin, the volume weighted mean pH of 14 rain events (Sept. 1980 to March 1981) was 4.81 (Galloway et al., 1982). Excess-SO_4 and NO_3 did not seem to be the main contributions to the free proton concentration.

4. ACIDITY OF GLACIERS AND POLAR ICE SHEETS

4.1. High Arctic

Koerner and Fisher (1982) reported acid concentrations in snow and ice samples collected on Northern Ellesmere Island (81°N, 78°W) and representing the last 26 years (snow samples) or several millenia (ice samples). pH was determined in meltwater. All samples were found to be acidic, with pH definitely lower than 5.6. The mean pH of recent snow was 5.28 for pit samples and 5.23 for core samples. The mean pH of Holocene ice (0-5000 yr BP) was found to be significantly higher : 5.48 (Figure 4). This increase

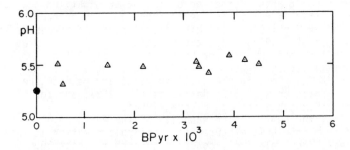

Figure 4 : The background acidity of High Arctic precipitation during the Holocene time period. Present-day mean pH value is also given (●). From data by Koerner and Fisher (1982).

of acidity (from 5.48 to 5.23) was interpreted by the authors as evidence of the recent pollution of the High Arctic by the long range transport of anthropogenic acid products. Over the last 26 years (Figure 5), a general

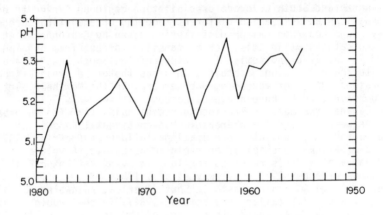

<u>Figure 5</u> : The increasing acidity of High Arctic snow during the last
26 years (from Koerner and Fisher, 1982).

decrease of 0.007 pH unit/year was calculated. These results are consistent
with the finding of anthropogenic sulfates at Barrow, Alaska (Rahn and Mc
Caffrey, 1979, 1980) at Bear Island (Rahn et al., 1980) in the Norwegian
Sea and in Arctic Canada (Barrie et al.,1981). Regular and well marked
seasonal variations were observed in the pH depth-profile, the acidity
being at a maximum in spring. pH oscillations were generally in the range
5.0-5.4. No influence of explosive volcanic activity, such as in Greenland
(see paragraph 4.2), was evident in these measurements.
      The spring maximum is most likely due to the advection of polluted
air masses into the polar regions in confirmation with an enhanced transformation of pollutants at the end of the polar night. On the contrary, in
summer the Arctic is relatively well protected from anthropogenic influences
since the polar front lies, in this season, north of the industrialized
areas (Rahn et al., 1977). Recently pH values as low as 4.7 were found in
Arctic pack ice, but the average values for precipitation sampled during
a cruise at latitudes higher than 78°N between Greenland and Franz Joseph
Land was 5.12 (Winkler, in press). This value agrees with the pH reported
for spring Arctic snow (5-5.2, Barrie et al., 1981) and for Ellesmere
Island snow (5.28, Koerner and Fisher, 1982).
      Finally, the neutralizing capacity of the North Polar atmosphere
estimated from insoluble measurements particulate matter in snow and ice,
seems to have been relatively constant through the past 5000 yr (Hammer,
1977 ; Koerner and Fisher, 1982).

4.2. Greenland

      Greenland snow and ice sampling stations are generally located at
lower latitudes and at higher altitudes than the Arctic sites reportes in
the preceding paragraph. The meteorological conditions prevailing in
Greenland are also not necessarily the same as those in the Arctic.
      No data concerning the direct determination of the pH of Greenland

snow are presently available, most acidity depth profiles having been
obtained using electroconductivity (Hammer, 1977, 1980 ; Hammer et al.,
1980). The relatively high pH values (6.0 to 7.5) found by Berner et al.
(1978) and used sometimes as a reference for the pre-industrial level of
snow acidity in Greenland were most likely due to contamination. The
background acidity value (i.e. during times of low volcanic activity)
reported by Hammer et al., (1980) is 1.2 μEq. H^+ l^{-1} at Crête (71°N, 37°W)
which corresponds to a pH of about 5.4 for meltwater in equilibrium with
ambient air at 0° C. It was assumed by these authors that acidities significantly above this background level were due to volcanic acid fallout.
    It is interesting to note that no recent acidity increase, such as
the one observed in the Canadian High Arctic, and attributable anthropogenic pollution is found when examining the 1850-1950 acidity depth
profile in the firn at Crête (Hammer et al.,1980). On the other hand,
Herron (1982) does find an increase of $SO_4^=$ (x 3-4) and NO_3^- (x 2-2.5)
deposition in Greenland snow from 1900 up to now. The corresponding
increase of acidity would be ~ 3 μEq. H^+ l^{-1} at Dye 3 (65°N, 43°W).
This apparent discrepancy could tentatively be explained 1) by analytical
problems (see paragraph 2) and/or by the difficulty of comparing acidity
data obtained by different techniques 3) by the near complete neutralization by NH_3 of the anthropogenic sulfate contribution deposited in Greenland. The former explanation seems to be more likely than the latter,
since long-range transport of sulfate occurs as H_2SO_4 rather than neutral
sulfate (Barrie et al., 1981). Moreover NH_4^+ levels have not changed in
Greenland snow since the seventeenth century (Busenberg and Langway,
1979). The most recent papers of Koerner and Fisher (1982) and Herron
(1982) seem thus to have demonstrated that Greenland and High Arctic snow
chemistry is presently influenced by anthropogenic activity, a conclusion
which was not evident a few years ago (Boutron and Delmas, 1980).
    The main purpose of Hammer's work was to estimate the amount of acids
injected annually into the atmosphere by volcanic eruptions and to reconstruct the chronology of the great volcanic events of the past. Hammer was
the first to demonstrate that major volcanic eruptions of the Northern
Hemisphere are recorded in Greenland ice (Hammer, 1977). Exceptional high
acidity values (up to 10 μEq. H^+ l^{-1}) were obtained after the Laki eruption
in Iceland (1783). More generally the acidity increase following an eruption rarely exceeds 2 μEq. H^+ l^{-1} one or two years after the date of
eruptions. The contribution of volcanoes to Greenland snow acidity might
be due to a long range tropospheric transport of acid products (mainly
H_2SO_4, but sometimes HCl or HF) rather than to stratospheric fallout,
except in the case of a few exceptional eruptions like Tambora (1815).
    Hammer et al. (1980) suggested that periods of frequent and violent
eruptions (i.e. periods of relatively high acidity level) usually coincided
with cold climatic conditions (for instance the Little Ice Age), but the
volcanic origin of all acidity fluctuations in Greenland snow was not
clearly demonstrated. In fact, the problem of the source of background
acidity is similar in all areas of the high Northern latitudes. It is
interesting to note that the NO_3^- determinations achieved by Hammer et al.
(1980) on a limited number of ice cores indicated that an important
part of the total acidity could be due to HNO_3 (but NO_3^- is not necessarily
bounded to H^+). Herron (1982) found also background NO_3^- concentrations
of ~ 1 μEq. l^{-1}. The origin of this compound in the Greenland
atmosphere is unknown· Herron rejects the assumption (Parker et al., 1978)
that polar NO_3 concentrations could be connected to solar activity).
    Finally the analysis of the Camp Century ice core in North-west

Greenland revealed that ice from the last glaciation was slightly alkaline (Hammer et al., 1980). It was proposed that wind blown $CaCO_3$ dust from continental margins neutralized the atmosphere acidity during this period.

4.3. Antarctica

The Antarctic atmosphere is much more remote from major continental and industrial aerosol sources than all other places in the world. The marine aerosol may therefore be considered to be dominant except at the most central locations of the Antarctic plateau. Delmas et al. (1982) pointed out that mineral acids are most probably the main impurities in Central Antarctic snow and ice. The quantitative and qualitative evaluation of this acidity has been done quite recently.

4.3.1. Coastal regions. A detailed study was achieved by Aristarain (Aristarain, 1980 ; Aristarain et al., in press) on 15 yr of precipitation deposited on the James Ross Island (Antarctic Peninsula, 64°S, 56°W, elevation 1600 m). The results of acidity measurements are similar to those found by Koerner and Fisher (1982) on Northern Ellesmere Island : the H^+ depth profile (Figure 6) exhibits strong seasonal variations (maximum in

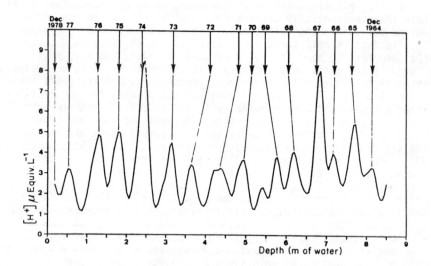

Figure 6 : The acidity of snow on James Ross Island (Antarctic Peninsula) (from Aristarain et al., in press).
The depth profile (14 years recorded) exhibits strong annual variations.

summer) in the range 1 to 7 μEq. l^{-1} (with the exception of 2 yearly peaks up to 12 μEq. l^{-1}). The average H^+ concentration (without taking into account carbonic acid) is 3.1 μEq. l^{-1} for the time period 1964-1978.

The strong correlation (0.84 < r < 0.92) found between "excess sulfate" (non sea-salt sulfate) and acidity shows that acidity is mainly contributed by H_2SO_4. Nitrate concentrations are relatively low (0.4 µEq. 1^{-1}, average value). Aristarain's results suggest that photochemical processes occuring only during the polar day could be responsible of the H_2SO_4 summer maximum.

The origin of H_2SO_4 in the Antarctic and Subantarctic atmosphere is not entirely explained. Nguyen et al. (1974) detected significant amounts of SO_2 over the subantarctic ocean.

It was suggested that, as it is the case at other latitudes, the biogenic activity releases substantial amounts of gaseous sulfur compounds (such as dimethylsulfide) which should be oxidized into SO_2 and acids SO_4. The absence of known large marine NH_3 sources explains that acid SO_4 is incorporated into snow without being neutralized. This assumption is supported by measurements in Adelie Land snow showing that ammonium concentrations are indeed very low (< 0.3 µEq. 1^{-1}, Legrand, to be published). Finally no definite trend which could be linked to anthropogenic acid pollution appeared in the James Ross acidity profile over the last 25 years.

4.3.2. Central Antarctic Plateau. The acidity of central Antarctic snow was investigated only recently : shallow firn layers analyzed from pit samples ; ancient ice layers (up to 30,000 yr) in ice cores. Several hundreds of samples were measured (Legrand, 1980 ; Delmas et al., 1980 ; Delmas et al., 1982 a).All of them were found to be acidic, the H^+ concentrations ranging generally between 1 and 5 µEq. 1^{-1} (corresponding pH of meltwater 5.2-5.5). Seasonal variations are more difficult to observe than in Greenland or than at James Ross Island due to the very low accumulation rate of snow (< 10 g cm^{-2} a^{-1}).

No definite trend which could be attributed to an increase of anthropogenic acid fallout in the recent decades can be found at Dome C (last 100 years profile). The probably most original finding from central Antarctic snow and ice acidity measurements is the impact of stratospheric volcanic H_2SO_4 on the sulfur cycle in these areas (Delmas and Boutron, 1980). Major S-producing volcanic events occuring south of 20° N may give a signal in snow : peak value was 6-7 µEq. H^+ 1^{-1} for Agung eruption, up to 14 µEq. 1^{-1} for the most important eruption detected in the Dome C ice core, 23,000 yr ago. It seems therefore that a record of the composition in the past of the stratospheric sulfate layer (Junge layer) exists in the Antarctic ice sheet. Based on the last 100 yr profile, the volcanic contribution to the total H_2SO_4 fallout on the Antarctic Plateau was estimated to be $\sim$ 30 % (Delmas and Boutron, 1980 ; Delmas et al., 1982 b).

Contributors to baseline acidity are most probably the biogenic activity of the Antarctic Ocean (as already suggested) and unknown atmospheric sources of HNO_3. The part of acid nitrate to the total acidity seems to be less than at lower latitudes. Potential sources and mechanisms of formation of Antarctic HNO_3 were reviewed by Parker et al. (1982). However concentration levels formely published by these authors are not in agreement with the data of Herron (1982) and Delmas et al. (1982 b).

A special mention must be made of the work by Millar (1982) who radioechosounded the Greenland and Antarctic ice sheets in order to detect acid layers recorded in the ice. This method, still at an experimental stage is assumed to give an indication of the volcanic acid fallout over the last 150 Ka. From such acidity profiles in Antarctica, Millar found a "low" of volcanicity during the most intense part of the last glaciation (15-20 Ka BP). Antarctic ice acidity (in fact the envelope of high acidity values) was found to oscillate between 5 and 10 µEq. H^+ 1^{-1} (Millar, 1982).

4.4. Alpine glacier

Reliable data concerning the acidity of alpine ice are very scarce.
Chemical data concerning temperate glaciers (glaciers at 0° C) are hard to
interpret because percolation water washed out soluble constituents. Even
for "cold" glaciers (those having mean temperatures below 0° C) summer
melting may cause considerable changes in the chemistry of the deposited
snow. Moreover, difficulties in dating the firn or ice layers in such
glaciers limits the conclusions which may be drawn. Delmas and Aristarain
(1978) analyzed firn samples collected near Mt Blanc, in an area without
significant summer melting. The acidity profile exhibited near neutral
values in pre 1963 firn. On the other hand, the firn collected in a pit
and corresponding to the years 1973-76, was found to be clearly acidic
($\sim$ 10 μEq. 1^{-1}). These measurements need to be confirmed because of the
large scatter in the values. Lyons et al. (in press) reports, for surface
Himalaya snows, lower pH at higher elevations. In the Canadian Yukon
(Mt Logan, elevation $\sim$ 5000 m) preliminary results (Legrand, personal communication) show that the acidity of 25 year old snow was $\sim$ 2-3 μEq. H⁺
1^{-1}.

5. DISCUSSION

A summary of precipitation acidity data in the selected background
regions described in chapters 3 and 4 are presented in the Figure 7. It
will be discussed in detail in relation to our present day knowledge of
atmospheric chemistry.

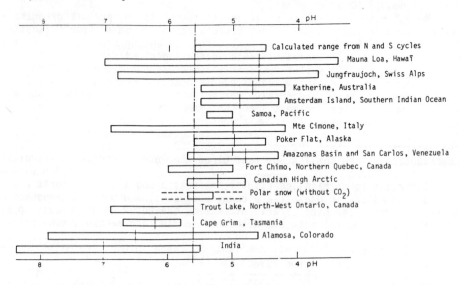

Figure 7 : The pH values of background precipitation collected in various
marine, continental or polar sites (data from different
authors, see text). 5.6 is the pH of pure water in equilibrium
with atmospheric CO_2. The pH of polar snow may be more acid
than indicated (after a strong volcanic eruption) or more
alkaline (during the last glaciation, see text).

5.1. The formation of acid precipitation in the atmosphere

The composition of solid and liquid precipitation is the final result of various processes within the atmosphere. From the time when the cloud elements are formed until the precipitation reaches the ground, trace substances can be incorporated into the condensed water-phase. In the atmosphere, no homogeneous phase transition for the water vapor occurs because suitable particles are always present, even at the lowest supersaturations, to initiate and facilitate a heterogeneous transfer of H_2O molecules to the condensed phase. The chemical composition of cloud elements therefore strongly depends on the composition of cloud active nuclei. To understand the acidity (or alkalinity) in background precipitation it is therefore necessary to look into the chemical composition of remote aerosols.

The acid-alkaline properties of aerosols were first determined by Junge and Scheich (1969). They found that the particles of the smallest size range, sampled in rural areas, reacted acidic whereas the coarsest fraction reacted more alkaline. This tendancy was also discovered to be valid for marine aerosols present 15 m above the North Atlantic (Gravenhorst, 1975 a), and in the Arctic (Winkler, 1980, in press) as well as in an industrialized area in Germany (Müller, 1982). This feature can be explained by the fact that the coarsest aerosol is dominated by particles (primary aerosol) which are emitted by sea bubbling or blown up from frequently basic mineral soils whereas the smallest particles are formed by gas-to-particle conversions leading to acidic nuclei (secondary aerosol). The gaseous precursors of these nuclei are generally anhydrids of acids (e.g. SO_2, N_2O_5) or even acids (e.g. HCl). The secondary aerosol over the oceans as well as in remote continental areas is largely characterized by sulfur compounds (Georgii and Gravenhorst, 1972 ; Meszaros and Vissy, 1974 ; Winkler, 1975 ; Gravenhorst, 1975 a, b).

On a global scale, alkaline gaseous ammonia cannot apparently totally neutralize the acidic sulfur components in the small particle size range (Gravenhorst, 1978). Despite the extensive discussion of the equilibrium relation between aerosol droplets and gaseous ammonia and nitric acid (Lee and Brosset, 1979 ; Tang, 1980) it is not possible at the moment to predict the exact acidic-alkaline properties of gas-derived aerosols since they are also controlled by kinetics of chemical and diffusional processes in the atmosphere. The observed acidic reactions of secondary aerosol when dissolved in water suggest that cloud droplets formed on these particles should also be acidic.

The coarsest particles derived from the sea and the soils contain alkaline components (sea-salt and carbonates) which tend to increase the pH of the water in which they are incorporated. Sea-water has a pH value of approximately 8.2. The sea-salt concentration in rainwater is, however, smaller than in ocean water by a factor of about 10^{-3} -10^{-4} so that the buffering capacity of sea-water is nearly exceeded and the alkaline influence largely reduced (Gravenhorst, 1975 a ; Pszenny et al., 1982). The presence of mineral soil components in the atmosphere is quite apparent in arid regions. Mineral soil particles can, however, be transported over the oceans (Prospero and Carlson, 1972 ; Gravenhorst, 1975 a ; Schütz, 1980) from one continent to another (Rahn et al., 1979) and even reach the high Arctic and Subarctic latitudes (Rahn et al., 1979 ; Petit et al., 1981 ; Briat et al., 1982). Their influence on the acidity of precipitation will be most pronounced near their source area. Whether the bulk water of precipitation samples will be acidic or alkaline depends on the relative contri-

bution of both aerosol size ranges to the total aerosol content of rainwater. The coarse aerosol mode measured at the ground is, however, not
the same as the one incorporated in cloud and rain droplets, because mixing
ratios of coarse particles tend to decrease with increasing altitude. This
is due to their settling velocity which cannot be neglected in comparison
to the turbulent transfer. Small particles which can also be formed aloft
have a minimal settling velocity so that they are usually distributed
more homogeneously in surface and cloud layers than the coarse particles.
Such different profiles have been found for small sulfate containing
particles and for sea-salt particles in maritime regions (Dinger et al.,
1970 ; Hobbs, 1971 ; Gravenhorst, 1975 a).

The incorporation of aerosol in cloud rain droplets depends on the
hygroscopie nature and the physical properties of the particles. It can be
assumed that small cloud droplets are more acidic than large ones. Since
sampling devices tend to collect the small droplets less efficiently than
the bigger ones (Fricke et al., 1978), the sampled water will probably
not represent the bulk cloud water. The large variations of pH values
reported for cloud water in remote areas (Oddie, 1962 ; Petrenchuk and
Drozdova, 1966 ; Lazrus et al., 1970 ; Scott, 1978 ; Hegg and Hobbs, 1981)
therefore provides no clear and unanimous interpretation. In arid areas
and under strong sea spray influence, rain should be alkaline. In areas
where the soil is protected against wind erosion by vegetation and where
sea-salt contribution is small or normal, precipitation should be acidic.
These features seem to be represented in the geographical distribution of
precipitation pH-values over the Northern Hemisphere (Gravenhorst et al.,
1980 ; Georgii, 1982). In the arid regions of South-western United States
and over the Indian subcontinent, relatively high pH values (pH > 6)
occur (see paragraph 3.4.). The low pH values of 4.0-4.5 in North-western
United States, Canada and Central Europe are, however, not only the result
of reduced alkalinity due to reduced sea and soil derived minerals in
precipitation, but also certainly include an effect related to anthropogenic acid forming emissions.

The Amazonian basin is mostly covered by humid tropical forest. Eolian
dust therefore does not dominate and rainwater seems to be acidic. In
marine regions, precipitation acidity seems to increase as the sea-salt
incorporated in the droplets decreases. Atmospheric transport of continental trace substances may either increase (Bermudas, see paragraph 3.3)
or decrease (Cape Grim) the acidity. In Antarctic regions, snow is ultra-
pure and therefore not buffered. Small traces of gas derived aerosols
(paragraph 4.3) make its meltwater acidic. This is enhanced when volcanic
traces are incorporated or reduced when small minaral soil particles are
deposited on the ice sheet (as in the case of Greenland during the last
glaciation). The general pattern of the geographical distribution of free
acidity in precipitation can thus be estimated knowing the main characteristics of the aerosol composition.

5.2. The importance of trace gases

Rainwater composition can nevertheless be influenced not only by the
aerosol phase, but also directly by the gas-phase. Trace gases which are
only physically dissolved in droplets such as CH_4, N_2O and H_2 do not change
the acid-base relation in droplets. If, however, dissolved species such
as H_2O-SO_2 or H_2O-CO_2 dissociate into ions and undergo further reactions,
the free acidity of the droplets changes. Aqueous phase reactions of
gaseous trace-compounds compete, therefore, with the dissolved aerosol in
relation to the free acidity of rainwater. Their respective influence

depends mainly on the gas-phase concentration, the reactant concentration
in the droplets, the pH value and the duration of interaction.

Far from their sources, reactive gaseous species are progressively
transformed into aerosol so that, for remote regions where no large point
sources of trace gases are present, it can generally be suggested that
the aerosol phase will override the gas-phase influence on rainwater
chemistry.
Among the gaseous sulphur species COS , CS $_2$, H_2S, $(CH_3)_2S$ and SO_2, only
the SO_2 interaction with the droplet phase seems to be of importance
(Gravenhorst, 1982). Although the SO_2-concentration is smaller in remote
than in more densely populated areas by a factor of about 10 (according
to a survey of SO_2-measurements in various parts of the world, Janssen-
Schmidt et al., 1981), the amount of SO_2 reversibly absorbed in droplets
may be the same in both regions (Gravenhorst et al., 1980). When going
from industrialized to remote areas, the decrease of the SO_2 concentrations
in the air may be counterbalanced by the decrease of the water droplet
acidity. Using the chemistry of SO_2 in solution, it was predicted (Barrie,
1981) that at low temperatures and with airborne particles howing a low
H^+ content, the reversibly absorbed SO_2 can constitute substantially to
the acidification of rain. 15 % of the free protons in precipitation in
Germany could be due to absorbed gaseous SO_2 (Gravenhorst et al., 1980) but
we are not aware of such measurements in remote areas.

The oxidation rate of SO_2 in the water droplets depends on a variety
of parameters (e.g. the production rate and the diffusion of reactants,
the catalyst and inhibitor concentrations, the droplet surface area, the
depletion of the gas phase) so that a general rate expression cannot yet
be given for comparison with acidification by aerosol incorporation.

For NO_2, a clear picture seems to be emerging at the moment, indicating that NO_2 does not greatly affect the droplet free acidity directly
(Lee and Schwartz, 1981, 1982 ; Gravenhorst, 1982). This compound leads
rather, via homogeneous gas phase reactions, to other nitrogen oxides
which in turn can contribute to aerosol and droplet acidification (Kessler
et al., 1982 ; Gravenhorst, 1982). Whereas for sulfate, the production
mechanisms seem to be better understood in the atmosphere than in the water
droplets, it seems to be the opposite for nitrate. Quantitative rate
expressions for nitrate formation in aerosols from gaseous precursors are
yet not available.

For the formation of ammonium in aerosols, only a first approach has
been proposed for central Europe (Lenhard and Gravenhorst, 1980). At the
moment it is therefore impossible to assess, in a reaction kinetic approach,
the contribution of gaseous and particulate phases to the free acidity
in background precipitation. A first attempt, however, to synthetize homogeneous gas-phase and aerosol-phase reactions of sulfur and nitrogen
compounds with aqueous phase interactions has been made using a one dimensional steady state model (Gravenhorst, 1982).

5.3. The prediction of background precipitation acidity

Another way to assess the acidity in cloud and rainwater in remote
areas is to calculate the reversible equilibrium distribution between the
gas and the liquid phase, assuming specific values for the concentrations
of trace gases and the liquid water content and assuming kinetic data for
the transformation of absorbed species. Such an approach was parameterized
for a system of SO_2, NH_3, CO_2 and liquid water by Scott and Hobbs (1967)
and by Barrie et al (1974). The resulting pH-values cover a wide range
depending on the assumed input parameters. They generally fall below 5.6.

The contribution of absorbed SO_2 to the ac idification of rainwater was discussed by Barrie (1981). According to his estimation, rainwater should have a pH value between 4.6 and 5.2 when 1 $\mu g/m^3$ acidic aerosol sulfate is present in the air and the SO_2 concentration falls between 0 and 10 $\mu g/m^3$. Barrie suggested that the best approach to predict a pH value of rain was to model the sulfate cycle and exclude the cycles of nitrogen oxides and ammonia until they are better known. This suggestion was apparently followed by Charlson and Rodhe (1982). Their approximation, which however, does not take into account the oxidation of absorbed SO_2, supports the overall validity of the older estimations.

The chemical composition of cloud-active nuclei determines so a great extend the pH-value of atmospheric water droplets, not only because of their own dissolution but also because of their effect on the uptake rate of gaseous compounds and on the transformation kinetics of absorbed species.

To estimate the natural acidity of cloud and precipitation water, we shall follow Barrie's (1981) idea and try to set up the natural cycles of trace compounds which determine the acid-base properties of atmospheric precipitation and to estimate the relative contribution of each species to the rainwater free acidity. The considerations of the effect of the natural sulfur cycle (Charlson and Rodhe, 1982) are therefore extended to include the atmospheric cycles of nitrogen oxides and ammonia.

The global production rates of natural sulfur and nitrogen compounds on which our estimation is based are given in Table 2. We assume that about half of the NO_x emitted and produced within the atmosphere returns to the ground in rain. This figure is based on the evaluation of Böttger et al. (1978), resulting in a ratio of about 1:1 between dry and wet deposition of total (natural and anthropogenic) NO_x emissions. For ammonia the ratio is about 1:2 indicating that relatively more ammonia ends up as ammonium in rainwater that NO_x as nitrate. The amount of wet deposited natural NO_x ranges therefore between 4 and 12 Tg (as N) a^{-1}. This value reaches 14 to 20 Tg a^{-1} for ammonia (from Table 2). The NH_3 production rate is, however, dominated by the decomposition of excrements from domestic animals so that the NH_3 emission rate used, although certainly of biological origin, is the result of partially man-induced processes. Taking this into account, it seems that the nitrate and ammonium flux from the atmosphere to the ground, derived from natural NO_x and NH_3 emissions, are of similar magnitude. With respect to the free adicity of rainwater, it can be said that, globally, the two parts of the N-cycles should roughly counterbalance each other. The worldwide bulk deposition rates of total ammonium and total nitrate including products of all sources reflect this balance (Böttger et al., 1978).

The molar ratio R of NH_4^+ and NO_3^- in bulk deposition was found to be roughly equal to unity, with higher NH_4^+ deposition rates at higher latitudes $(1 < R < 2)$ than at lower latitudes $(0.5 < R < 1)$. This "neutralization" of alkaline and acid nitrogen compounds in the global precipitation does not necessarily mean that NH_4NO_3 is formed in the atmosphere. Remote aerosol measurements revealed that NH_4^+ and NO_3^- are linked to two quite distinct or even separate particle modes : NH_4^+ on the smaller particles and nitrate on the larger particles (Gravenhörst et al., 1979 ; Mamame and Pueschel, 1980).

Although ammonia and nitrogen oxides have quite different sources and path ways in the atmosphere, they seem to end up in similar amounts in precipitation. Rainwater sampling, storage and analyzes for ammonium still, however, represents a severe problem (Hayes et al., 1980 ; Müller et al., 1982).

Source	Production rate $(Tg\ a^{-1}$ of S or N)	Reference
Sulfur		
Volcanoes	7	Berresheim et al. in press
Vegetation fires	1	Janssen-Schmidt et al. (1981)
$(CH_3)_2S$	3-34	Nguyen et al. (1978) Barnard et al. (1981) Varhelyi and Gravenhorst (1981)
H_2S	0.05-1.8)	
COS	0.07-0.3 (	" " "
CS_2	0.2-1.8 (	" " "
Total	11-46	
Nitrogen oxides		
Lightning	1-4	Dawson (1980), Hill et al. (1980)
Bushfires	0.9-2.1	Böttger et al. (1978)
Soil	5-10	Galbally and Roy (1978)
Ocean	0.0007	Gravenhorst and Böttger (1983)
NH_3 oxidation	1-5	Böttger et al. (1978)
N_2O destruction	1-2	" " "
Total	9-23	
Ammonia		
Decomposition of organic matter in soil	1-2	Gravenhorst and Böttger (1980)
Decomposition of animal excrements	20-30	Böttger et al. (1978)
Total	21-32	

Table 2 : Global amounts of S and N compounds emitted into the atmosphere

If the alkaline ammonium and the acid products of gaseous nitrogen oxides balance each other in rainwater, the amount of sulfur species in precipitation water could represent the amount of free acid formed when no other alkaline components (e.g. soil dust) are present. Of the total natural sulfur emission of 30 $\pm$ 20 Tg a^{-1} (Varhelyi and Gravenhorst, 1981), about 20 $\pm$ 15 Tg a^{-1} find its way into precipitation water. The value (2/3) of this fraction is based on the world wide ratio of 186 $\pm$ 41 Tg a^{-1} (as S) for wet and dry deposition and 135 $\pm$ 39 Tg a^{-1} for wet deposition of sulfur compounds neglecting sea-salt sulfur (Janssen-Schmidt et al., 1981).

The effect of this estimated amount of 20 $\pm$ 15 Tg S a^{-1} incorporated in rainwater and coming from natural gaseous sulfur precursors will lead to an acidification of the natural precipitation. The resulting pH-value of this precipitation water will then depend on the alkaline contribution of atmospheric bases (besides ammonia).

The geographical distribution of nitrogen oxide, sulfur compound and ammonia natural sources are quite different. NH_3 sources are confined to the ground in continental areas (Söderlund and Svensson, 1976 ; Böttger et al., 1978 ; Gravenhorst and Böttger, 1980). Twelve potential NO_x sources were reviewed recently by Parker et al. (1982). This trace gas can be formed in the stratosphere via reactions of N_2O with atomic oxygen. This stratospheric NO_x source should give a wide spread nitrate flux in rain. Lightning can form NO_x in upper layers of the troposphere even over the sea, so that not only the horizontal but also the vertical distribution of NO_x and NH_3 sources differ greatly. Moreover, biogenic sulfur sources are distributed in quite another way. In contrast to nitrogen oxides and ammonia, natural sulfur seems to volatilize from the oceans in considerable amounts (Nguyen et al., 1978 ; Barnard et al., 1981). Natural SO_2 seems to be directly emitted into the atmosphere only from volcanoes, which means its emission height is distributed over the first 5 km of the troposphere (Simkin et al., 1981 ; Gravenhorst, 1982), if single events of large volcanic eruptions reaching into the stratosphere are neglected. The conversion of COS to acidic sulfate proceeds preferentially in the lower stratosphere . These CS_2 can also produce acidity when it is converted via the COS pathway. These different distributions in the atmosphere for the natural emission and conversion of nitrogen oxides, ammonia and gaseous sulfur species underline that the global flux considerations will not be totally fulfilled for each specific situation in space and time. The source and conversion distribution, however, in general explains our picture of the aerosol and precipitation chemistry in remote areas. The higher the altitudes in which the measurements are performed, the more acidic the samples react. In the middle troposphere, the alkaline effect of ammonia, soil dust and sea-salt is reduced so that acidic compounds of sulfur and nitrogen oxides dominate. This tendency was suggested by Gravenhorst (1977) when discussing aerosol measurements above and below the trade wind inversions over the oceans and is further supported by model calculations of the interaction of gaseous and condensed nitrogen and sulfur compounds with the droplet phase (Gravenhorst, 1982).

The rather uniform aerosol, which is found in altitudes above the marine and continental aerosol and was characterized by Junge (1963) on the basis of Aitken nuclei concentrations and further described by Jaenicke (1980), should therefore react acidic. Cloud and rainwater formed in this region should consequently also be acidic.
The same tendancy seems to be valid when going from continental to marine regions. An unique region seems to be Central Antarctica. Because of its elevation and remoteness from continental and marine sources, reference as well as background levels of atmospheric trace compounds may be found for

past and present conditions.

The background acidity of precipitation depends strongly on conditions at and above the sampling site. These are not only affected by the local environment, but also by the advection of natural and anthropogenic influences from large distances. Since these parameters can vary quite extensively from place to place and from time to time, the free acidity in background precipitation can vary considerably. The range of background pH values suggested in Table 3 for rainwater in remote regions of the world is therefore only a first attempt aimed at generalizing the results of the rather variable composition of background atmospheric precipitation. Despite this variability, free acidity levels as high as those found in rainwater in industrialized regions has not yet be found in remote areas.

Region	pH-range
Continental	
vegetation covered soil	4.5 - 5.5
arid soil	5.8 - 7.4
Maritime	
low elevation	4.8 - 5.8
high elevation	4.0 - 5.5
Polar	
Arctic and Greenland background	5.0 - 5.5
Natural reference	5.3 - 5.6
Antarctic background and natural reference	5.2 - 5.7 *

Table 3 : Tentative classification of background precipitation and its pH values (in equilibrium with atmospheric CO_2, except *).

ACKNOWLEDGMENTS

We wish to thank Dr. Sterzky (Eidgenössisches Materialprüfungsamt, Dübendorf, Switzerland) for providing the unpublished pH values of Jungfraujoch reported in the Figure 7.

REFERENCES

Aristarain, A.J., 1980, Etude glaciologique de la calotte polaire de l'île James Ross (Péninsule Antarctique). Thèse de 3e cycle, Laboratoire de Glaciologie, Publ. N° 322, 130 p.

Aristarain A.J., R.J. Delmas, and M. Briat, in press. Snow chemistry on James Ross Island (Antarctic Peninsula). J. Geophys. Res.

Askne, C., and C. Brosset, 1972; Determination of strong acid in precipitation, lake water, and air-borne matter, Atm. Environment, 6, 695-

Barnard, W.P., M.O. Andrae, W.E. Watkins, H. Bingemer, and H.W. Georgii, 1981, The flux of demethylsulfide from the oceans to the atmosphere, IAMAP Symposium "the role of oceans in atmospheric chemistry", Hamburg, August 1981.

Barrie, L., S. Beilke, and H.W. Georgii, 1974, SO_2 removal by cloud and rain drops as affected by ammonia and heavy metals, Proc. Precip. Scavenging Symp., Champaign, (Ill., USA), Oct. 14-18.

Barrie, L.A., 1981, The prediction of rain acidity and SO_2 scavenging in Eastern North America, Atm. Environment, 15, 31-41.

Barrie, L.A., R.M. Hoff, and S.M. Daggupaty, 1981, The influence of mid latitudinal pollution sources on haze in the Canadian Arctic, Atm. Environment, 15, 1407-1419.

Berner, W., B. Stauffer, and H. Oeschger, 1978, Past atmospheric composition and climate, gas parameters measured on ice cores, Nature, 276, 53-55.

Berresheim, H., and W. Jaeschke, in press. The contribution of volcanoes to the global atmospheric sulfur budget. J. Geophys. Res.

Blanchard, D.C., and A.H. Woodcock, 1980, The production concentration and vertical distribution of sea-salt aerosol, in : Aerosols : anthropogenic and natural, sources and transport, The New York Acad. of Sciences, New-York (T.J. Kneip and P.J. Lioy Eds) pp 330-347.

Bonsang, B., 1982, Le soufre dans l'atmosphère, La Recherche, 13, 1132-1142.

Boutron, C., and R. Delmas, 1980, Historical record of global atmospheric pollution revealed in polar ice sheets, Ambio, 9, 211-215.

Böttger, A., H.D. Ehhalt, and G. Gravenhorst, 1978, Atmosphärische Kreisläufe von Stickoxiden und Ammoniak, Berichte der KFA Jülich, Nr. 1558.

Brezonik, P.L., E.S. Edgerton, and C.D. Hendry, 1980, Acid precipitation and sulfate deposition in Florida, Science, 208, 1027-1029.

Briat, M., A. Royer, J.R. Petit, and C. Lorius, 1982, Late glacial input of eolian continental dust in the Dome C ice core : additional evidence from individual microparticle analysis, Annals of Glaciol., 3, 27-31.

Busenberg, E., and C.C. Langway, Jr., 1979, Levels of ammonium, sulfate, chloride, calcium and sodium in snow and ice from Southern Greenland, J. Geoph. Res., 84, 1705-1709.

Charlson, R.J., and H. Rodhe, 1982, Factors controlling the acidity of natural rainwater, Nature, 295, 683-685.

Dawson, G.A., 1980, Nitrogen fixation by lightning, J. Atm. Sci., 37, 174-178.

Delmas, R., and A. Aristarain, 1978, Recent evolution of strong acidity at Mt Blanc in : Studies in Environmental Science, M.M. Benarie (Ed.), 1, 233-237.

Delmas, R., and C. Boutron, 1980, Are the past variations of the stratospheric sulfate burden recorded in central Antarctic snow and ice layers ? J. Geophys. Res., 85, 5645-5649.

Delmas, R., A. Aristarain, and M. Legrand, 1980, The acidity of polar precipitation : a natural reference level for acid rains, in : Ecological impact of acid precipitation, Proc. of an Intern. Conf. Sandefjord, Norway, March 11-14, 1980, Drabløs and Tollan (Eds), pp 104-105.

Delmas, R., M. Briat, and M. Legrand, 1982 a, Chemistry of South Polar snow, J. Geophys. Res., 87, 4314-4318.

Delmas, R.J., J.M. Barnola, and M. Legrand, 1982 b, Gas derived aerosol in cintral Antarctic snow and ice : the case of sulphuric and nitric acids, Annals of Glaciol., 3, 71-76.

Dinger, J.E., H.B. Howell, and T.A. Wojciechowski, 1970, On the source and composition of cloud nuclei in a subsidient air mass over the North Atlantic, J. Atm. Sci., 27, 791-797.

Fanning, C.D., and L. Lyles, 1964, Salt concentrations of rainfall and shallow ground-water across the lower Rio Grande Valley. J. Geophys. Res., 69, (4), 599-604.

Fricke, W., H.W. Georgii, and G. Gravenhorst, 1978, Application of a new sampling device for cloud water analysis, in : Some problems of cloud physics (collected papers). Gidrometeoizdat, Leningrad (USSR), 200-212.

Galbally, I.E., and C.R. Roy, 1978, Loss of fixed nitrogen from soils by nitric oxide exhalation, Nature, 275, 734-735.

Galloway, J.M., G.E. Likens, W.C. Keene, and J.M. Miller, 1982, The composition of precipitation in remote areas of the world, J. Geophys. Res., 87, 8771-8786.

Georgii, H.W., and G. Gravenhorst, 1972, Untersuchungen zur konstitution des Aerosols über dem Atlantischen Ozean, Met. Rdsch., 25, 180-181.

Georgii, H.W., 1982, Global distribution of the acidity in precipitation, in : Deposition of atmospheric pollutants, H.W. Georgii and J. Pankrath (Eds), D. Reidel Publ. Comp., Dordrecht, Boston, London, PP 55-56.

Government of India, 1982, Atmospheric turbidity and precipitation chemistry data from background air pollution monitoring stations in India (1973-1980) issued by : The director general of Meteorology, India Meteorological Dpt.

Gran, G., 1952, Determination of the equivalence point in potentiometric titrations. Part. II. Analyst, 77, 661-671.

Granat, L., 1972, On the relationship between pH and the chemical composition in atmospheric precipitation, Tellus, 24, 550-560.

Granat, L., H. Rodhe, and R.O. Hallberg, 1976, The global sulphur cycle, in : Nitrogen, Phosphorus and Sulphur-Global Cycles (B.M. Svensson, and R. Söderlund Eds), SCOPE Rep. 7, Ecol. Bull. (Stockholm), 102-122.

Grant, M.C., and W.M. Lewis Jr., 1982, Chemical loading rates from precipitation in the Colorado Rockies, Tellus, 34, 74-88.

Gravenhorst, G., 1975 a, Der Sulfatanteil im atmospherischen Aerosol über dem Nordatlantik, Berichte des Instituts für Meteorologie und Geophysik der Universität Frankfurt am Main, Nr. 30.

Gravenhorst, G., 1975 b, The sulfate component in aerosol samples over the North Atlantic, "Meteor" Forsch. Ergeb. Ber., 10, 22-31.

Gravenhorst, G., 1977, Marine aerosol in : Atmospheric Aerosols and Nuclei, A.F. Roddy and T.C. O'Connor (Eds), Galway University Press, pp. 468-476.

Gravenhorst, G., 1978, Maritime sulfate over the North Atlantic, Atm.Environment, 12, 707-713.

Gravenhorst, G., K.P. Müller, and H. Franken, 1979, Inorganic nitrogen over the North Atlantic, Gesellschaft für Aerosolforschung, 7, 182-187.

Gravenhorst G., and A. Böttger, 1980, Ammonia in the atmosphere, First European Symposium : Physical-chemical behaviour of atmospheric pollutants. Ispra, 16-18 Oct. 1979, B. Versino and H. Ott (Eds), 383-395.

Gravenhorst, G., S. Beilke, M. Betz, and H.W. Georgii, 1980, Sulfur dioxide absorbed in rainwater, in : Effects of acid precipitation on terrestrial ecosystems, Hutchison and Havas (Eds), Plenum Publ. Corp., pp. 41-55.

Gravenhorst, G., 1982, Der Einfluss von Wolken und Niederschlag auf die vertikale Verteilung atmosphärischer Spurenstoffe in einem eindimensionalen reaktionskinetischen Modell, unpubl. Rep. Laboratoire de Glaciologie, Grenoble.

Gravenhorst, G., and A. Böttger, 1983, Field measurements of NO and NO_2 fluxes to and from the ground, this volume.

Hammer, C.U., 1977, Past volcanism revealed by Greenland ice sheet impurities, Nature, 270, 482-486.

Hammer, C.U., 1980, Acidity of polar ice cores in relation to absolute dating, past volcanism and radio-echoes, J. of Glaciol., 25, 359-372.

Hammer, C.U., H.B. Clausen, and W. Dansgaard, 1980, Greenland ice sheet evidence of past-glacial volcanism and its climatic impact, Nature, 288, 230-235.

Harding, D., and J.M. Miller, 1982, The influence on rain chemistry of the Hawaiian volcano Kilauea. J. Geophys. Res., 87, 1225-1230.

Hayes, D., K. Snetsinger, J. Ferry, V. Oberbeck and N. Farlow, 1980, Reactivity of stratospheric aerosols to small amounts of ammonia in the laboratory environment, Geophys. Res. Letters, 7, 974-976.

Hegg, D.A. and P.V. Hobbs, 1981, Cloud water chemistry and the products of sulfates in clouds, Atm. Environment, 15, 1597-1604.

Herron, M.M., 1982, Impurity sources of F^-, Cl^-, NO_3^-, and SO_4^{--} in Greenland and Antarctic precipitation, J. Geophys. Res., 87, (C4), 3052-3060.

Hill, R.D., R.G. Rinkler, and H.D. Wilson, 1980, Atmospheric nitrogen fixation by lightning, J. Atm. Sci., 37, 179-192.

Hobbs, P.V., 1971, Simultaneous ariborne measurements of cloud condensation nuclei and sodium containing particles over the ocean, Quart. J.R. Met. Soc., 97, 263-271.

Janssen-Schmidt, T., E.P. Röth, G. Varhelyi, and G. Gravenhorst, 1981, Anthropogene Anteile am atmosphärischen Schwefel-und Stickstoff-Kreislauf und mögliche globale Auswirkungen auf chemische Umsetzungen in der Atmosphäre, Bericht K.F.A., Jülich, Nr. 1722.

Jaenicke, R., 1980, Atmospheric aerosols and global climate, J. Aerosol. Sci., 11, 577-588.

Jickells, T., A. Knap, T. Church, J.N. Galloway, and J. Miller, 1982, Acid rain on Bermuda, Nature, 297, 55-57.

Junge, C.H., 1963 a, Air chemistry and Radioactivity, Academic Press, New-York, London.

Junge, C.H., 1963 b, Large scale distribution of condensation nuclei in the troposphere, J. Rech. Atm., 1, (4), 185-189.

Junge, C.H. and G. Scheich, 1969, Studien zur Bestimmung des Säuregehalts von Aerosolteilchen, Atm. Environment, 3, 423-441.

Kessler, C., D. Perner, and U. Platt, 1982, Spectroscopic measurements of nitrons acid and formaldehyde - implication for urban photochemistry, in : Physico-chemical behaviour of atmospheric pollutants, Proc. 2d. Eur. Symp. Varese, Sept. 29-Oct. 1, 1981, pp. 393-400.

Khemani, L.T., and B.H.V. Rama Murty, 1968, Chemical composition of rainwater and rain characteristics at Delhi, Tellus, 20, 284-292.

Klockow, D., H. Denzinger, and G. Rönicke, 1978, Zum Zusammenhang zwischen
   pH-Werte und Elektrolytzusammensetzung von Niederschlagen, VDI Berichte
   314, 21-26.
Koerner, R.M., and D. Fisher, 1982, Acid snow in the Canadian High Arctic,
   Nature, 295, 137-140.
Lazrus, A.L., H.W. Baynton, and J.P. Lodge, 1970, Trace constituents in
   oceanic cloud water and their origin, Tellus, 22, 106-114.
Lee, Y.H., and C. Brosset, 1979, Interaction of gases with sulphuric acid
   aerosol in the atmosphere, in : WMO Symposium on the long range transport of pollutants and its relation to general circulation including
   stratospheric-tropospheric exchange processes, Sofia, Bulgaria,
   1-5 Oct., 1979.
Lee, Y.H., and S.E. Schwartz, 1981, Reaction kinetics of nitrogen dioxide
   with liquid water at low partial pressure, J. Phys. Chem. 85, 840-848.
Lee, Y.H., and S.E. Schwartz, 1982, Evaluation of the rate of uptake of
   nitrogen dioxide by atmospheric and surface liquid water, J. Geophys.
   Res., in press.
Legrand, M., 1980, Mesure de l'acidité et de la conductivité électrique
   des précipitations antarctiques. Thèse de 3e cycle, Grenoble,
   Publ. N° 316 du Laboratoire de Glaciologie.
Legrand, M.R., A.J. Aristarain, and R.J. Delmas, 1982, Acid titration of
   polar snow, Anal. Chem. 54, 1336-1339.
Lenhard, U., and G. Gravenhorst, 1980, Evaluation of ammonia fluxes into the
   free atmosphere over western Germany, Tellus, 32, 48-55.
Liberti, A., M. Possanzini, 1972, M. Vicedomini, 1972, The determination of
   the non volatile acidity of rainwater by a coulometric procedure,
   Analyst, 97, 352-356.
Lyons, W.B., P.A. Mayewski, and N. Ahmad, in press, Acidity of recent
   Himalayan snow, 38th Eastern Snow Conf.
Mamame, Y., and R.F. Pueschel, 1980, A method for the detection of indivi-
   dual nitrate particles, Atm. Environment, 14, (6), 629-639.
Meszaros, A., and K. Vissy, 1974, Concentration, size distribution and
   chemical nature of atmospheric aerosol particles in remote oceanic
   areas, J. Aerosol Sci., 5, 101-109.
Millar, D.H.M., 1981, Radioecho layering in polar ice sheets and past vol-
   canic activity, Nature, 292, 441-443.
Millar, D.H.M., 1982, Acidity levels in ice sheets from radio echo-sounding,
   Annals of Glaciol., 3, 227-232.
Miller, J.M., 1980, The acidity of Hawaiian precipitation as evidence of
   long range transport of pollutants, WMO Publ. N° 538, 231-237.
Miller, J.M., and A.M. Yoshinaga, 1981, The pH of Hawaiian precipitation,
   a preliminary report, Geophys. Res. Letters, 8, (7), 779-782.
Müller, J., 1982, Residence time and deposition of particle bound atmosphe-
   ric substances in : Deposition of Atmospheric Pollutants H.W. Georgii
   and J. Pankrath (Eds) D. Reidel Publ. Comp. pp. 43-52.
Müller, K.P., G. Aheimer, and G. Gravenhorst, 1982, The influence of immediate freezing on the chemical composition of rain samples in :
   Deposition of atmospheric pollutants, D. Reidel Publ. Comp.,
   H.W. Georgii and J. Pankrath (Eds), pp. 125-132.
Nguyen, B.C., B. Bonsang, and G. Lambert, 1974, The atmospheric concentration of sulfur dioxide and sulfate aerosols over Antarctic, Subantarctic areas and oceans, Tellus, 26, 241-249.
Nguyen, B.C., A. Gaudry, B. Bonsang, and G. Lambert, 1978, Reevaluation of
   the role of dimethyl sulphide in the sulphur budget, Nature, 275,
   637-639.

Oddie, B.C.V., 1962, The chemical composition of precipitation at cloud
    levels, Quart. J.R. Met. Soc., 88, 535-538.
Parker, B.C., L.E. Heiskell, W.J. Thompson, and E.J. Zeller, 1978, Non
    biogenic fixed nitrogen in Antarctica and some ecological implications,
    Nature, 271, 651-652.
Parker, B.C., E.D. Zeller, and A.J. Gow, 1982, Nitrate fluctuations in
    antarctic snow and firn : potential sources and mechanisms of forma-
    tion, Annals of Glaciol., 3, 243-248.
Petit, J.R., M. Briat, and A. Royer, 1981, Ice age aerosol content from
    East Antarctica ice core samples and past wind strength, Nature, 293,
    391-394.
Petrenchuk, O.P., and V.M. Drozdova, 1966, On the chemical composition on
    cloud water, Tellus, 18, 280-286.

Prospero, J.M., and T.N. Carlson, 1972, Vertical and areal distribution
    of Saharan dust over the western equatorial North Atlantic ocean,
    J. Geophys. Res., 77, 5255-5265.
Pszenny, A.A.P., F. Mac Intyre, and R.A. Duce, 1982, Sea-salt and the
    acidity of marine rain on the windward coast of Samoa, Geophys. Res.
    Letters, 9, (7), 751-754.
Rahn, K.A., R.D. Borys, and G.E. Shaw, 1977, Particulate air pollution
    in the Arctic : large scale occurence and meteorological controls,
    in : Proc. 9th Intern. Conf. on Atm. aerosols, condensation and ice
    nuclei, Galway, Ireland, 21-27 Sept. 1977, pp. 223-227.
Rahn, K.A., and R.J. Mc Caffrey, 1979, Compositional differences between
    Arctic aerosol and snow, Nature, 280, 479-480.
Rahn, K.A., R.D. Borys, G.E. Shaw, L. Schütz, R. Jaenicke, 1979, Long range
    impact of desert aerosol on atmospheric chemistry. Two examples, in :
    Saharan dust : Mobilisation, Transport, Deposition, SCOPE Rep. 14,
    C. Morales Ed. Wiley and Sons, Chichester, UK, pp. 243-266.
Rahn, K.A., and R.J. Mc Caffrey, 1980, On the origin and transport of the
    winter Arctic aerosol, in : Aerosols : Anthropogenic and natural,
    sources and transport, Annals N.Y., Acad. Sci., 338, 486-503.
Rahn, K.A., E. Joranger, A. Semb, and T.J. Conway, 1980, High winter concen-
    tration of SO$_2$ in the Norwegian Arctic and transport from Eurasia,
    Nature, 287, 824-826.
Schütz, L., 1980, Long range transport of desert dust with special emphasis
    on the Sahara, Amer. New-York Acad. Sci., 338, 515-532.
Scott, R.V. and P.V. Hobbs, 1967, The formation of sulfate in water droplets,
    J. Atm. Sci., 24, 54-57.
Scott, W.D., 1978, The pH of cloud water and the production of sulfate,
    Atm. Environment, 12, 917-924.
Sequeira, R., 1981, Acid rain : some preliminary results from global data
    analysis, Geophys. Res. Letters, 8, 147-150.
Sequeira, R., 1982, Acid rain : an assessment based on acid-base considera-
    tions, J. Air Poll. Contr. Ass., 32 (3), 241-245.
Simkin, T., L. Siebert, L. Mc Clelland, D. Bridge, C. Newhall, and J.H.
    Latter, 1981, Volcanoes of the world, Hutchinson Ross Publ. Comp.,
    Strondsburg, Pensyl. USA.
Söderlund, R., and B.H. Svensson, 1976, The global nitrogen cycle, in :
    N, P, S - cycles, B.H. Svensson and R. Söderlund (Eds), SCOPE Report 7,
    Ecol. Bull. (Stockholm), 22, 23-73.
Stallard, R.F., and Edmond J.M., 1981, Geochemistry of the Amazon : 1. Pre-
    cipitation chemistry and the marine contribution to the dissolved load
    at the time of peak discharge. J. Geophys. Res., 86, (C 10), 9844-9858.

Tanaka, S., M. Darzi, and J.W. Winchester, 1980, Sulfur associated elements and acidity in continental and marine rain from North Florida, J. Geophys. Res., 85, 4519-4526.

Tang, I.N., 1980, On the equilibrium partial pressures of nitric acid and ammonia in the atmosphere, Atm. Environment, 14, 819-828.

Tyree, S.Y.Jr., 1981, Rainwater acidity measurement problems. Atm. Environment, 5, 57-60.

Varhelyi, G., and Gravenhorst, 1981, An attempt to estimate biogenic sulfur emission into the atmosphere, Időjaras, 85, (3), 126-133.

Winkler, P., 1975, Chemical analysis of Aitken particles (< 0.2 μm radius) over the Atlantic ocean, Geophys. Res. Letters, 2, 45-48.

Winkler, P., 1980, Observations on acidity in continental and in marine atmospheric aerosols and in precipitation. J. Geophys. Res., 85, (C8), 4481-4486.

Winkler, P., in press, Acidity of aerosol particles and of precipitation in the north polar region and over the Atlantic, Tellus.

TRENDS IN THE ACIDITY OF RAIN IN EUROPE: A RE-EXAMINATION
OF EUROPEAN ATMOSPHERIC CHEMISTRY NETWORK DATA

A.S. KALLEND
Central Electricity Generating Board,
Leatherhead Laboratories,
Kelvin Avenue, Leatherhead,
Surrey, United Kingdom

Summary

Data from the European Atmospheric Chemistry Network over the period
1956-76 has been evaluated statistically to establish what trends
are apparent in precipitation acidity. Out of 120 sites with 5 or
more years data 29 show a significant trend of increasing annual
average acidity in precipitation during the period and 5 show a
decrease. For these sites substantially the same result arises for
the hydrogen ion concentration calculated from ionic balance based on
detailed chemical analysis of the precipitation samples. Examination
of the monthly data shows that the increased annual average levels for
H^+, where they do occur, arise from an increased frequency of intermittent high monthly values. At several stations this appears to
have started quite suddenly around 1965 leading to an apparent step
change in annual average acidity. Data from most of the sites does
not appear to fit the simple picture of an expanding area receiving
acid precipitation over the period in question.

1. INTRODUCTION

One of the major questions in the current acid rain debate concerns
the extent which precipitation has become more acid during the last few
decades and how such trends, if they do exist, match changes in the emissions of the major pollutants which are the precursors to acidity.

By far the most extensive body of information on precipitation chemistry in Europe comes from the European Air Chemistry Network (EACN), also
called the IMI Network after the International Meteorological Institute in
Stockholm which operated it. The Network originated in about 1947 when a
number of precipitation stations were established in Sweden in connection
with a study of the transport of nitrogen compounds as part of an agricultural research programme on nutrient cycles. The scope of the programme
was subsequently widened to include other chemical components and by 1955
sampling stations were established in eleven other European countries.
The number of stations grew to a maximum of about 120 in 1959 but decreased to about 50 by the late 1970's.

To begin with equipment for all sampling sites was supplied from Sweden
and early analysis were performed at a laboratory in Uppsala. Several
other laboratories became involved by the late 1950's and samples from
Belgium, Netherlands and France were analysed in Belgium from 1957; Irish
samples in Ireland from 1958; British samples in Britain from 1959 and
Finnish samples in Finland from 1961. The rest of the Network analyses and
data from all stations were handled at the special laboratory set up in

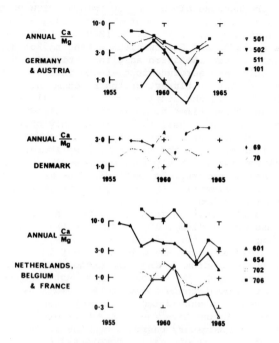

**FIG.1 TRENDS IN Ca/Mg RATIO
AT EUROPEAN STATIONS 1956-65
(AFTER PATTERSON, 1975)**

1963 at the International Meteorological Institute, University of
Stockholm.

The EACN data consists of monthly measurements of precipitation amount
in millimetres, its pH, the deposition of ions, SO_4 (as sulphur) Cl, NO_3
and NH_4 (as nitrogen) Na, k, Mg and Ca in mg m^{-2} along with the concentration of HCO_3^- in microequivalent l^{-1} and the conductivity in micro cm^{-1}.
The collectors were continuously opened and so collected both wet and dry
deposited material.

Previous publications, principally by Granat (1972, 1978), have focuss-
ed principally on the sulphate content of precipitation and, surprisingly,
in spite of the current interest in acid rain, no complete analysis has
been presented of the precipitation acidity data derived from the EACN,
although the data for individual stations has been published from time to
time (Oden 1968, 1976, Likens et al 1979) and Granat presented data on deposition of net acid for the period up to 1970 (Granat 1972).

This paper summarises an evaluation of the EACN data on precipitation acidity which is presented in more detail elsewhere (Kallend et al, 1982). The
purpose is to attempt to establish what trends are apparent for the period
1955-1975. For comparison data is also presented for both sulphate and
nitrate concentrations.

2. RESULTS

It is not the purpose of this paper to deal with possible systematic and other errors arising, for example, from inadequacies in analytical protocol. Granat (1972a, 1972b, 1975, 1978), Paterson and Scorer (1973, 1975), Paterson (1975) and Galloway and Likens (1976) have discussed sources of error in precipitation sampling, storage and analysis. For example, the siting and aerodynamic characteristics of precipitation gauges has an important influence on their contents. In this context it should be noted that in the mid 1960's the original EACN collectors were replaced at least three times by modified designs (Granat 1977). All of these collector types were used to obtain monthly bulk samples.

Analytical methods for several components changed as improved techniques became available. With the move to a new laboratory in Stockholm in the mid 1960's several new analytical methods were introduced in the main (Swedish) analytical centre. In particular the method for sulphate was changed from the colorimetric barium sulphate procedure to the present-day coulometric thorin method.

Some systematic errors in methodology were noted by Paterson (1975) and Paterson and Scorer (1973, 1975). They observed, for example, that in the case of samples from three Austrian stations, a change of analytical laboratory in 1964 was accompanied by a three-fold increase in both calcium and magnesium concentrations. The same authors also pointed out the discrepancy between annual Ca/Mg ratios in different European countries. A strong pattern of variation is common to all stations in Germany and Austria yet it does not extend the relatively short distance into Belgium France or Denmark (Figure 1). The factor determining the ratio thus appeared to be the laboratory carrying out the analysis.

Chemical data from precipitation chemistry network is customarily presented as volume-weighted annual averages and so accordingly the initial analysis in the present study was carried out in this way. Annual average concentrations were evaluated for each chemical species and, in the initial analysis, no data points were rejected although a large number of stations had data missing for one or two months in a given year or data missing for two or three months for just a few of the ions, notable NH_4. In those instances the values for the missing months were assumed to correspond to the yearly mean values. Where less than 6 months data were available yearly means were not evaluated. On the basis that the minimum of five years data was necessary to evaluate time trend, suitable data sets were available for 120 stations on the network. For each station a linear regression analysis was carried out and the correlation coefficient between H^+ concentration and time evaluated over the period for which data were available. The detailed results have been published elsewhere (Kallend, Marsh, Pickles and Proctor 1982).

A statistically significant positive trend in concentration is shown by 29 out of the 120 stations for H^+, 23 for sulphate and 55 for nitrate. On the other hand, five stations showed a significant decrease in H^+, one for sulphate and none for nitrate. Ten stations showed a significant trend for both sulphate and H^+, and 18 for both nitrate and H^+. The trend in nitrate concentration has been noted previously both in relation to EACN data (OECD 1978) and in North America (Likens et al 1977).

For those sites showing trends based on the weighted annual average data the regression analyses were repeated using monthly data but eliminating data for those months where the ionic balance was worse than ± 10%. This reduced the number of sites showing a significant positive trend in H^+ from 29 to 24 and those showing a significant decrease in H^+ concentration from 5 to 4.

Cogbill and Liken (1974) suggested that it should be possible to evaluate pH even when direct measurements are not available, if a complete analysis of the remaining ionic components exists. On this basis H^+ was evaluated for all those sites showing a positive trend in measured H^+ concentrations. 21 of the sites showed a significant positive trend in H^+ concentration when the 10% ion balance filter was imposed upon the data and none showed a significant trend.

For about one third of the stations showing a significant trend in H^+ concentration, a more or less sudden change in precipitation acidity seems to have occurred in the mid 1960's rather than a gradual change over the whole time period. Figure 2 shows three examples for Scandinavian stations. This behaviour mirrors similar observations with respect to both sulphate and concentration (Figure 2) reported by Granat (1978) and could be related to a sudden, but maintained, increase in annual deposition of sulphate. Alternatively it could reflect the known changes in sampling and analytical techniques referred to above which took place at about that time.

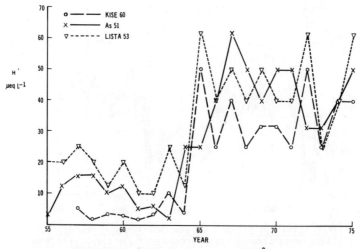

FIG. 2 ANNUAL MEAN H⁺ CONCENTRATIONS v. YEAR FOR LISTA 53, Ås 51 AND KISE 60

Twelve sites, all of them in Scandinavia, had sufficient data available for separate regressions to be run for two individual time periods namely (i) up to the end of 1964 and (ii) from January 1965 onwards. Eight of these sites showed a significant positive trend over the whole time period but the rate of increase in measured H^+ in the post-1965 period does not attain statistical significance (p < 0.05) for any individual station. Neither this behaviour nor that for sulphate illustrated, for example, by the data in Figure 3 seems to reflect the increase in total European SO_2 emission of more than 30% that occurred over this latter time period.

Kallend et al (1982) attempted to examine the data from sites in the same general geographical area to seek correlations in their analytical data. Four sites around Uppsala all within 25 km of each other were examined. They were: Uppsala (43), Bjorsund (47), Ryda Kungsagrd (42) and Tarna (48). It was noted that the former pair of sites both showed a significant negative trend of H^+ concentration with time whilst the latter two showed a non-significant but positive trend. For each pair of stations the correlation indicated that there was a detectable common regional component in the variations data which accounted for 20-30% of the variants. The greater part of the variance, however, was accounted for by measurement factors particular to each individual station.

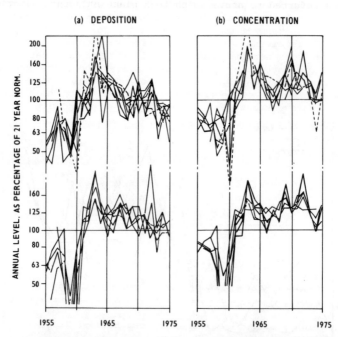

FIG. 3 EXCESS SULPHATE DEPOSITION (a) AND CONCENTRATION (b) AT TWO GROUPS OF SCANDINAVIAN STATIONS OF THE EUROPEAN ATMOSPHERIC CHEMISTRY NETWORK. AFTER GRANAT, (1978. 1977)

3. DISCUSSION AND CONCLUSIONS

A simple regression analysis on the weighted annual average precipitation acidity of sites in the European and Atmospheric Chemistry Network shows that only 29 out of 120 sampling sites gave a statistically significant trends of increasing acidity over 5 or more years in the period 1955-1975. Data filtering on the basis of ionic balance of each individual monthly sample does not materially alter this conclusion so that, for these sites, the hydrogen ion concentrations in the 1970's are higher, typically by a factor by 3 or 4 than the concentrations measured in the 1950's. The increase in nitrate concentration over the period, however, is particularly noticeable as has been remarked upon elsewhere. A detailed analysis does not bear out the expectation of a steady increase in acidity over the whole period of 1955-1975 paralleling, for example, an increase in sulphate emissions. Rather, the monthly data show that the increased annual average levels for H^+, where they do occur, arise from an increased frequency of intermittent high monthly values. At several stations this seems to have started quite suddenly around 1965 leading to an apparent step change in annual average acidity.

Bearing in mind the kind if experimental inadequacy demonstrated by Paterson and Scorer as well as by others, interpretation of the changes in precipitation chemistry should clearly proceed with care. Although the data from the sites showing a trend of increasing acidity broadly fit the isopleth maps (Likens et al, 1979) which are commonly taken to show an expanding area receiving acid precipitation in Europe, the data from most of the sites does not fit that pattern. The isopleth maps fail to adequately show the pattern of time variation over the years and the large degree of uncertainty evidently attaching to individual contours. They appear to assume a degree of geographical homogeneity which is not borne out by detailed calculations on data taken from adjacent sites.

4. REFERENCES

Cogbill, C.V. and Likens, G.E., (1974), Water Resources Res., 10, No. 6, 1133.
Galloway, J.J. and Likens, G.E., (1976), W.A.S.P., 6, 241.
Granat, L., (1972a), Tellus, 24, 550.
Granat, L., (1972b), I.M.I. Report AC-20.
Granat, L., (1975), I.M.I. Report AC-30.
Granat, L., (1977), I.M.I. Report AC-40.
Granat, L., (1978), Atmos. Env., 12, 413.
Kallend, A.S., Marsh, A.R.W., Pickles, J.H. and Proctor, M.V., (1982), Atmos. Env.m in press.
Likens, G.E., Bormann, F.H., Easton, J.S., Pierce, R.S. and Johnson, N.M., (1977), Bio-geochemistry of a Forested Ecosystem, Springer-Verlay, New York.
Likens, G.E., Wright, E.F., Galloway, J.N. and Butler, T.J., (1979), Scientific American, 241, No. 4, 39.
Oden, S., (1968), Swedish Nat. Sic. Res. Council, Ecology Committee, Bul. 1, 68.
O.E.C.D. (1978), Report on Long Range Transport of Air Pollution, O.E.C.D., Paris.
Paterson, M.P., (1975), Ph.D. Thesis, University of London.
Paterson, M.P. and Scorer, R.S., (1973), Atmos. Env. 7, 1163.
Paterson, M.P. and Scorer, R.S., (1975), Nature, 254, 491.

TREND DEVELOPMENT OF PRECIPITATION-PH

IN CENTRAL EUROPE

P. Winkler
Deutscher Wetterdienst
Meteorologisches Observatorium Hamburg
Frahmredder 95, D-2000 Hamburg 65

Summary

A critical investigation of literature reported precipitation-pH measurements leads to the conclusion that
since the late 30th the precipitation-pH has not changed
very much in Central Europe, in spite of the fact, that
the SO_2 emission has doubled in the meantime. This
finding is confirmed by some observations of the precipitation-pH in Hamburg, which is measured since 1976
with an automatic precipitation monitor, operating continuously during falling precipitation. The pH shows no
pronounced yearly cycle and no dependence of the wind
direction, even in those cases where a trajectory analysis proves direct transport of the air mass over the
North Sea without land contact. It is concluded, that
acid formation during the precipitation process is limited due to physical and chemical reasons and that on
the average, the removal capacity of the precipitation
for acidifying gases is exhausted, when an average pH-
value slightly above 4 is reached. As a consequence of
increasing SO_2 emission a spreading out of the area
with acid rain rather than a local decrease of the pH
is predicted.

1. INTRODUCTION

In 1872 R.Smith published a book entitled "Air and
Rain - The beginning of a chemical climatology". In the introduction of the chapter "Rain" he wrote:
"It becomes clear from the experiments that rain-water
in town districts, even a few miles distant from a
town, is not a pure water for drinking; and that, if it
could be got direct from the clouds in large quantities,
we must still resort to collecting it on the ground in
order to get it pure. The impurities of rain are completely removed by filtration through the soil; when
that is done, there is no more nauseous taste of oil or
of soot, and it becomes perfectly tranparent. The presence of free sulfuric acid in the air sufficiently explains the fading of colours in prints and dyed goods,
the rusting of metals and the rotting of blinds".

and a few sentences later:
> "I do not mean to say that all rain is acid; it is often found with so much ammonia in it as to overcome the acidity; but in general, I think, the acid prevails in the town".

Smith conducted a lot of experiments to study its trace substance content. He carried out his experiments very carefully and being a person of sound diffidence, he collected the rainwater in a platinum basin which he, to prevent all mistakes, kept red-hot for sometime prior to use. If we do interpret his figures for the amount of free acid in our present understanding pH-values of 4 in the London rain, for example, were quite common. It seems to be a fruitful task, to reexamine his methods and results.

Any pH-values of precipitation, reported in the literature, must be investigated very critically with respect to collection method, analytical method and averaging procedure. For example, we find data of open gauges emptied only once every four weeks and thus representing the total, wet and dry deposition. Or in order to obtain a mean pH-value, the individual pH-values were averaged arihmetically although this should not be done with logarithms. Many authors try to avoid this difficulty by presenting frequency distributions of pH-values but their information content is also limited because the most frequent pH-value must not be coupled with the rain events producing the highest rain amounts. The best way of averaging pH-values is to calculate the average depositions (i.e. H^+ concentration times precipitation amount) and convert it back to the average pH. This procedure represents the pH-value measured in a very large reservoir, in which many precipitation events of several years are collected.

SO_2 and NO_2, the main acidifying gases are produced by many various sources. In Central Europe, although there are areas with major sources, the concentration differences from the different sources have leveled out to certain degree, before the gases are absorbed in clouds to form acid precipitation. Also precipitation is connected with wind from different directions, which again causes a transport from different sources and mixing equilization of pH differences.

2. DATA

Fig 1 puts together the data available from the literature. The earliest measurements were made by Ernst in 1937/38 Bad Reinerz and Oberschreiberhau both some 150 km NE of Prague. The measurements were performed very careful. The average values as indicated are calculated as precipitation amount weighted averages. The data of Ernst (1938) were also published by Drischel (1940) but together wit the precipitation amount. The average pH of the events of a one-and-half year period was 4.17.

The next measurements were carried out by Harrassowitz (1956) who was locking for an impact of nuclear test explotions on the precipitation pH. From a short notice we can con-

clude that Harrassowitz was   aware of the difficulties during collection and measurement and that his measurements in
the Westerwald mountain region were carried out very careful.
Unfortunately only the frequency of pH values was published
with the most values occuring in the pH range 4.2-4.5.

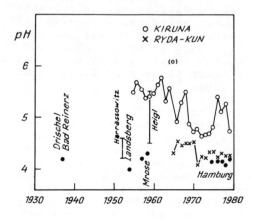

Fig 1
pH-values of
precipitation
during the last
5 decades from
various stations
in Central Europe

       Landsberg(1954) reported measurements of the pH of
single rain drops collected in Boston. We have taken his values for reasons of comparison. Boston is also an industrialized area and the values can be compared with European data,
as we will dicuss later, because the overall removal process
between different measuring sites seems to be very similar.
       The next measurements from the late 50th were published
by Mrose (1966)for the Wahnsdorf Observatory near Dresden.
His average value seems to be an arithmetic average of pH-
values as can be concluded by comparison with a frequency distribution. If we estimate an average value by weighting with
appropriate precipitation amounts (frequency distribution of
the precipitation amount is similar to that of the pH-values)
we obtain an average pH of 4.2, which seems to be more realistic.
       Heigel (1960) also reported measurements from the late
50th carried out at the Hohenpeissenberg mountain Observatory
south of Munich. Unfortunately he presented only a pH frequency distribution with a very broad class width of one pH unit.
Most values occured in the range between pH 4.5 and 5.5. The
main value may be slightly higher due to the mountain situation.
       Additional values exist from the network of the National
Science Foudation of FRG from 1967-1972 (Kaiser et al. 1974).
Because the published monthly pH-values are also arithmeti-
cally averaged valued they are not depicted. From the data
of the rural stations, however, it can be expected that a
mean value, weighted with the precipitation amount, would be

near 4.2.

The recent measurements of Hamburg, collected with the automatic precipitation analyzer of the Met. Obs. Hamburg (Winkler 1977, 1978) give a weighted average of pH 4.2. These values are consistent with the data collected by the network of the Environmental Protection Agency of FRG (UBA 1982) which demonstrate that there exist only minor regional differences of the pH-values from different stations.

These values taken alltogether, indicate that there has been no long-term trend of the pH-value at least during the last 50 years in Central Europe. Since the measurements suffer from many short comings and the reliability of the data can only by traced back in some cases, the statement of a missing pH trend must be supported by additional arguments.

3. ANALYSIS OF THE DATA OF HAMBURG

The data of a single station should also reflect the insensitivity of the pH-value to variable amounts of acidifying gases, as long as the concentration remains above a minimum level. To prove this, the data of Hamburg being collected with the automatic precipitation analyzer, are investigated. We begin with averaging of daily values of the electrical conductivities, which belong to same pH-value by weighting them with the precipitation amount and plotting these averaged conductivities into a pH-conductivity diagram. This averaging procedure corresponds to a hypothetic set of precipitation gauges, each labeled with a different pH-value and each opening and collecting precipitation only, if rain with its distinct pH is falling. The electrical conductivity which is measured in a gauge after a long time corresponds to the average value obtained by our averaging procedure. The pH-conductivity diagram contains a set of percentage-labeled inclined lines with the following meaning: If we take a pure diluted droplet of acid we measure a certain pH and a corresponding electrical conductivity. Concentration or dilution of the acid shifts the diagram point along the 100%-line. At a given pH adding of neutral reacting salts to the acid increases the conductivity letting the pH unchanged. Because the main ions dissolved in the precipitation remain always the same ones and because their average equivalent weight and average equivalent conductivity remain similar when the chemical composition is varied, we can calculate from the conductivity the amount of dissolved material after correction for the H^+ caused conductivity. In this way, the set of percentage labeled lines indicates which fraction of the dissolved matter consists of pure acid (Winkler 1980). 10% means that 10% of the total dissolved material consists of pure acid. The conductivity measured may not fall below a minimum value, which is defined by the conductivity of the H^+ ions alone. So the area above the 100%-line is a forbidden range and the diagram can be used for some quality control.

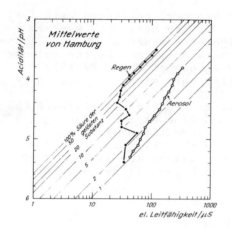

Fig 2
Simultaneously measured and averaged values of pH and electrical conductivity. Full dots: precipitation; open circles: aerosol particle leaching solutions. By comparison, it can be seen that aerosol acid contributes only a minor fraction to the acid found in the precipitation.

Fig 2 shows that at low pH-values the dissolved material is dominated by acid. For pH values above about 4.2 the acid fraction decreases very rapidly. For reasons of control we have shown similar measurements of aerosol samples, which were collected by an impactor. The samples were leached with several ml of water so that the aerosol of 1 m³ was dissolved in 0.4 ml of water. This amount is comparable to the liquid water content of clouds. It can be seen that the acid fraction of the aerosol usually ranges between 1 to 5% of the soluble matter while for precipitations of low pH it amounts to nearly 60%.

It is now most interesting that the precipitation collected at other sites exhibits a very similar characteristic relation between pH and electrical conductivity.(Fig 3 shows data of 3 Scandinavian stations). However,the pH above which the acid fractions lowers rapidly, is shifted to a lower average conductivity which can be interpreted in terms of reduced influence of the local pollution.

This charateristic relation between pH-value and electrical conductivity obviously describes the average equilibrium between the amount of trace substances available for rainout and washout in the atmosphere and the overall removal capacity of the precipitation process. If we imagine a pure airmass free of acidifying gases the acid fraction of the precipitation will reach a value determined by the aerosol. If we now add increasing amounts of SO_2, the acid fraction will at first increase very rapidly until the acid fraction reaches 50 to 60% of the total dissolved material. At this point the acid forming process has reached its maximum turn over capacity. This can be demonstrated by the fact that the **leg** of the characteristic relation (Fig 2 and 3) where the acid fraction increases, has a relatively constant electrical conductivity. Above that point further acidification proceeds more slowly, also in cases where the amount of acidifying gases increases.

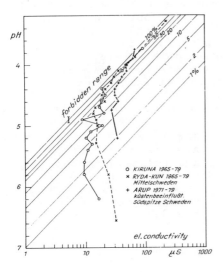

Fig 3
Simultaneous measure-
ments of pH and elec-
trical conductivity
for 3 Scandinavian
stations.

pH-values lower than 3.4 have never been observed and this value seems to be a lower limit, mainly because limitations of physical and chemical nature exist. At first life time of cloud or rain drops is limited and within this time only a limited amount of SO_2 (resp NO_2) can ce oxidized to sulfate (nitrate). Secondly, with decreasing pH dissolving of SO_2 becomes more difficult; also rate constants are pH dependent. pH-values lower than 3.4 sometimes reported in literature are doubtful.

Another important argument that a pH-trend in Central Europe is not very likely, can be derived from the air mass dependence of the pH-values. The daily average pH-values measured with the automatic precipitation analyzer were compared with the daily 850 mbar (48 h) trajectory. The trajectories were classified into 8 classes (Fig 4). For the trajectories falling into one sector pH-value and electrical conductivity were averaged by weighting them with the precipitation amount. As can be seen from Fig 4, the pH-values as well as the average acid fractions do not vary very much. It is very astonishing that in cases where the air had no land contact during transport directly over the North Sea, the acid fraction with 18% raises to only 24% when the air was transported over the Ruhr area. The difference between these two sectors should be enlarged by the local industry and the city of Hamburg.

The small air mass dependence, however, means that either acidification proceeds very rapidly even at low SO_2 concentrations or that the transport of pollutants is not described good enough, i.e. also no land contact is indicated, there exists a marked cross-trajectory flow below 850 mbar with continuous mixing upwards so that the SO_2 content of the clean air mass is higher than indicated by the trajectory transport over the North Sea. As already concluded from the discussion of the characteristic relation between pH and electrical conductivity, fast acidification even at low SO_2 concentrations seems to be the main reason for the small air mass dependence of the pH-values.

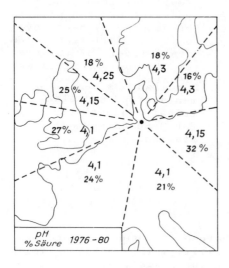

Fig 4
Average values of pH and relative amount of free acid for different air masses (48 h back-trajectories of the 850 mbar level).

This last result is confirmed by the fact that no yearly cicle is observed (Fig 5). Again, (6 years) average pH-values were calculated by weighting with the precipitation amounts. If we convert the pH to H^+ concentrations the yearly variation remains below a factor of two. For Hamburg we can expect the SO_2 emission to be in winter nearly twice as high as in summer. If there would be any propotionality between SO_2 amount and pH the yearly SO_2 variation should cause a pH variation of 0.3 units. Because in general the trace substance amount in rain water decreases as precipitation amount increases and highest precipitation amounts occur during the summer. The yearly pH cycle should be enlarged by another 0.3 pH units.

The absence of any clear pH variation through the year as well as the only slight dependence of the pH of the wind direction also clearly indicates that the formation of acid during the precipitation process is limited. It seems to be likely, that the removal capacity is exhausted, when an average pH-value of 4.2 is reached. Lower pH-value of 4.2 of course do occur in cases with favourable conditions for acid formation (small rain drops, little NH_3, plenty SO_2). A limited average pH-value, however, as deduced from the observations confirms our previous conclusion that we do not find a long-term pH trend. As a consequence of raising pollution, pH lowers at first until an average of 4.2 is reached. After that, increasing pollution does not lead to a further pH decrease, but the surplus of acidifying gases is transported away and acid precipitation is formed far from the sources.

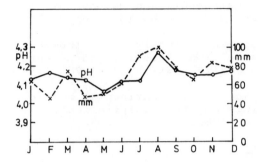

Fig 5
Yearly cycle of the
pH-value of precipitation
of Hamburg averaged over
6 years (full line). The
yearly cycle of the pre-
cipitation amount is in-
dicated by the broken-
line.

This is what was observed in Scandinavia (Oden 1976). In Central Europe, however, the point at which the air pollution
level was high enough to produce an average pH of 4.2 was
reached before 1935. In spite of the fact that the SO_2 emission has doubled since than, precipitation has not become
more acid but it has become also acid at remote places.

LITERATURE

Drischel, H. "Chlorid- Sulfat- und Nitratgehalt der atmosphä-
    rischen Niederschläge in Bad Reinerz und Oberschreiber-
    hau im Vergleich zu bisher bekannten Werten anderer Or-
    te". Der Balneologe 7: 321-334 (1940)
Ernst, W. "Über pH-Wertmessungen von Niederschlägen". Der
    Balneologe 5: 545-549 (1938)
Harrassowitz, H. "Atombombenexplosionen und Regen-pH". Die
    Naturwissenschaften 43: 11-12 (1956)
Heigel, K. "Die pH-Werte von Niederschlägen, Kondensation, Ne-
    belfrostablagerungen und der Schneedecke auf dem Hohen-
    peissenberg. Unterschiede der pH-Werte von Hohenpeissen-
    berg und Peissenberg". PAGEOPH 47: 142-154 (1960)
Kayser, K., U.Jessel, A.Köhler, G.Rönicke "pH-Werte des Nie-
    derschlags 1967-1972". Mitteilung der Kommission zur Er-
    forschung der Luftverunreinigung der Deutschen For-
    schungsgemeinschaft, Bold Verlag Boppard, 1974.
Landsberg, H. "Some observations of the pH precipitation ele-
    ments". Arch.Met.Geoph.Bioclim. Ser.7: 219-226 (1954)
Mrose, H. "Measurement of pH and chemical analysis of rain
    snow and fog-water". Tellus 18: 266-270 (1966)
Oden, S. "The acid problem - an outline of concepts". J.Air
    Water Soil Poll. 6: 137-166 (1976)
UBA (1982): Monatsberichte aus dem Meßnetz des Umweltbundes-
    amtes 7, No 1, Jan 1982. "Die zeitliche Entwicklung der
    überregionalen Pegel von Schwefeldioxid, Stickstoffdi-
    oxid und Schwefel im Schwebstaub und der Konzentration
    von H^+ Ionen im Niederschlag".

Winkler, P. "Automatic analyser for pH and electrical conduc-
    tivity of precipitation" in: Papers presented at the WMO
    Technical Conference on Instruments and Methods of Ob-
    servation(TECIMO) Hamburg 1977. WMO No 480, Geneva 1977,
    pp 191-196
Winkler, P. "Fehler bei der Spurenstoffanalyse im atmosphäri-
    schen Niederschlag dargestellt am Beispiel von pH-Wert
    und elektrischer Leitfähigkeit". Staub 38 (1978) 175-
    177
Winkler, P. "Observations on acidity in continental and in
    marine atmospheric aerosols and in precipitation". J.
    Geoph. Res. 85: 4481-4486 (1980)

ACID PRECIPITATION OVER THE NETHERLANDS

T.B. Ridder and A.J. Frantzen
Royal Netherlands Meteorological Institute
P.O. Box 201, 3730 AE De Bilt, The Netherlands

Summary

After some remarks about the history, information will be given about
networks measuring the chemical composition of precipitation. The
development of the networks is discussed. Siting of the stations is
one of the problems.
Up till now generally open (bulk) collectors have been used. This
results in contamination of the sample by birds' droppings. The difference between open and wet-only collectors is demonstrated by means
of some results of the analysis of rainwater obtained at the same site
with different types of collectors.
Into the acidification aspect of the precipitation will be looked more
closely.
The trend of the pH of rainwater during the period from 1956 to 1981
in The Netherlands will be discussed.

1.   From 1957 to 1974 the precipitation in the Netherlands was sampled
monthly at three stations. This was done in the framework of the European
Atmospheric Chemistry Network (EACN) organized by Sweden.
    In 1978 the Dutch network was extended to 12 stations and by joining
another network in 1983, there will be 20 stations taking monthly and
halfmonthly samples. This project is carried out by the Royal Netherlands
Meteorological Institute (KNMI) and the National Institute for Public
Health (RIV).
    Besides the main components also trace elements are analysed. (Table
1).

COMPONENTS	
Main Components	pH, conductivity H, NH_4, Na, K, Ca, Mg, Zn F, Cl, NO_3, SO_4, HCO_3, PO_4
Trace Elements	As, Cd, Co, Cr, Cu, Fe Mn, Ni, Pb, Se, V

Table 1.

2.   In our densely-populated country with much industry it is not easy to
find observing sites, which are not influenced by local sources. At this
moment we have 6 stations at airports and up till now we do not see any
influence on the main components. However, trace elements and organic micro
components may be influenced.

3.   Generally we use <u>open</u> collectors. Therefore we have made several pro-
visions to avoid birds' droppings, which could spoil the samples (fig. 1).

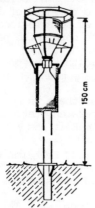

We find most of these droppings in the inland
during the summer months.

Besides these <u>open</u> collectors we have two opera-
tional wet-only collectors (type dr. Granat,
Sweden).

Fig. 1.

	Winter 1980/81 543 mm			Summer 1981 358 mm		
	open O	wet-only W	O–W / O %	open O	wet-only W	O–W / O %
NH4	88	62	18	104	90	13
Ca	9	6	33	15	9	40
NO3	41	36	12	64	59	8
SO4	55	43	15	69	57	17
pH	4,60	4,50	–28	4,33	4,30	–9

Difference in
concentration be-
tween an open
collector (O) and a
wet-only collector
(W) at De Bilt in
μmol/l.

Table 2.

    Table 2 shows the concentration data of some components in μmol/l for
a winter halfyear at De Bilt. (with the amount of precipitation). Under
"Open" the results are given of the open collector and next to it the results of the wet-only collector. Then the difference and the percentage of
this difference in ratio to the open collector.
    The same data are given for the summerperiod.
    Generally the concentration in the open collector is higher than in
the wet-only collector. That is what we expect. But in the open collector
the precipitation is less acid than in the wet-only one. We think this is
due to dry deposition of ammonium and calcium.
    Generally it can be said that the important acid making components
sulphate and nitrate at De Bilt for open and wet-only collectors show a
difference of less than 20%. Data from Witteveen confirm this.
    It may be that the differences between open and wet-only collectors
will be greater at stations nearby sources of pollution.

4.   Now we will have a look at the spatial distribution of some components.
      First <u>acidity</u> (pH).

This map (fig. 2) gives the mean acidity for 4 years. Plotted are the data of 13 stations. With the aid of some regional networks it was tried to detail the pattern of the isolines.
Generally it can be said that the precipitation is rather acid. The mean pH of De Bilt for 4 years is 4,3.

It can be seen that there is not much difference in acidity between stations. The two main stations De Bilt and Witteveen have nearly the same acidity.

Fig. 2. pH

5.   One of the important components for acidity is <u>Nitrate</u>. The average yearly Nitrate concentration at De Bilt is 53 μmol/l.

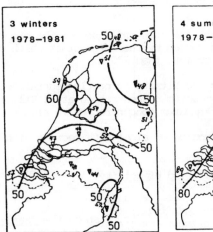
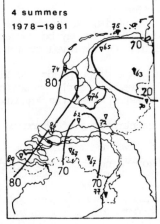

Fig. 3.   Nitrate in μmol/l

The maps (fig. 3) give the mean values for 3 winters and 4 summers. It can easily be seen that Nitrate in summer is higher than in winter. It may be that this is caused by photochemical reactions.
Generally in the Netherlands Nitrate provides about 30% of the acidity, but in some months up to 40% is due to this component.

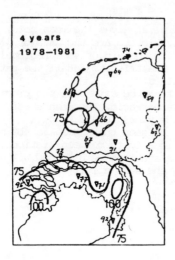

Fig. 4. Sulphate in μmol/l.

6.   About the _pattern of sulphate_ (fig. 4) is more to be said than about nitrate. Here we see the mean sulphate concentration for 4 years. The average yearly Sulphate concentration at De Bilt is 67 μmol/l.
     It looks like there are _2 main sources_
a) in the southeast
b) in the southwest.
     Another point is, we do not see much influence of the big oil refineries near Rotterdam. It may be caused by the large stack which discharged about 40% of all SO_2 at level of more than 200 m.

7.   The influence of Sulphate and Nitrate is partly neutralized by _Ammonium_. The average yearly Ammonium concentration at De Bilt is 98 μmol/l.
     In the coastal areas the Ammonium concentration is always less than in the inland.
     High concentrations of Ammonium go along with regions with much bio-industry.

8.   Lastly something about the _trend of acidity_.

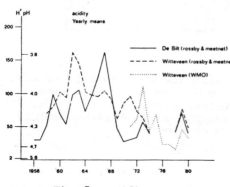

Fig. 5.   Acidity.

Fig. 5 shows the _yearly averages_ of the pH transferred to H-concentration for De Bilt, and Witteveen for the period 1956 to 1981.
     A high concentration of H means a low pH. For example in 1967 at De Bilt the yearly mean H-concentration was 166 μmol/l, that means a pH less than 3,8.
     It looks as if, the acidity was rising first and falling after about 1968. But there is much variation between De Bilt and Witteveen and you may remember this was not the case suring recent years.

- 126 -

Therefore we do not trust this picture completely.

In 1974 the measurements at the two stations stopped, but since 1971 another collector of the same type was installed at Witteveen. In 1978 the new network started and De Bilt and Witteveen produce more or less the same results.

During the years 1978, 1979 and 1980 <u>at Witteveen</u> two collectors were in operation:
the old one from 1971 onwards and
the new one from 1978 onwards.

The new one had very strict instructions for sampling, especially concerning birds' droppings. When we compare corresponding data from both collectors, then we see that the summerdata differ very much and the winter data are nearly the same. (Fig. 6).

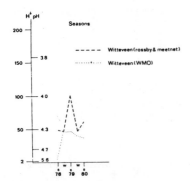

Fig. 6. Acidity.

This difference may be caused by contamination by birds' droppings, which, I told, most of the time occur in summer. Therefore it can be, that the old winter data are better than the summerdata.

<u>For the winter</u>
again we look at De Bilt and Witteveen (fig. 7).

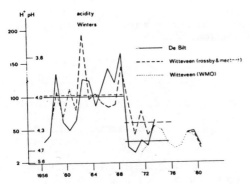

Fig. 7. Acidity.

The variation between stations is less than for the yearly data and further we see <u>a big decrease in acidity</u> after the winter 1968/69 both at De Bilt and at Witteveen. The mean of H-concentration in Witteveen suddenly drops from 100 to 60 and in De Bilt even from 100 to 33 μmol per liter. At Den Helder this phenomena does not occur. The winter mean overthere remains 70 μmol per liter.

Also the Belgian stations Ukkel and St. Andries show nearly the same picture as Den Helder with nearly the same mean before and after 1968.

I have the impression that something happened southwest of De Bilt and Witteveen, which did not influence acidity at Den Helder and the Belgian stations (fig. 8).

Fig. 8.

What can this be?
Around this year natural gas came in general use in the Netherlands, but it took several years to implement this.

In summer 1968 the oilindustry near Rotterdam brought a large stack into operation. Furthermore in summer 1969 low sulphur fuel was used in that part of the oil installations, that was not connected to the central stack. Also meteorological circumstances may have had influence too.

It may be several of these causes worked together. Up till now it is not possible to say which of these was the most important factor.

Acknowledgement.
Dr. H.F.R. Reijnders (RIV) is acknowledged for providing us with the data of chemical analysis of precipitation samples.

INVESTIGATION OF THE SOURCE REGIONS FOR ACID DEPOSITION IN THE NETHERLANDS

J. SLANINA[+], F.G. RÖMER[++] and W.A.H. ASMAN[+++]

[+]Netherlands Energy Research Foundation (ECN), Petten (N.H.)The Netherlands
[++]N.V. KEMA, KEMA Laboratories Environmental Department, Arnhem, The Netherlands, [+++]Institute for Meteorology and Oceanography (IMOU), Utrecht, The Netherlands.

Summary

To assess the impact on acid deposition of specific sources, information is needed about the background situation and about the contribution of individual sources.
The background situation (i.e. source regions, relative contribution of different strong acids etc.) is studied in an experiment performed by ECN and IMOU at Lelystad, in The Netherlands.
The specific contribution of a powerplant is investigated in a joint experiment of ECN, KEMA and IMOU near the Flevo powerplant. The methodology and some first results of this experiment are described.

INTRODUCTION

The impact of a single source of sulfur- and nitrogenoxides on the acidification of precipitation can only be assessed in comparison with the background situation i.e. the contribution of other source regions.
The characterization of the background situation should involve the following questions:
- Which acids contribute to the excess acidity in precipitation?
- Where are the source regions for these acids?
These questions cannot be answered directly due to a number of complications;
- It is difficult to evaluate the accuracy of sampling-, sample pretreatment- and analytical procedures used in investigations of precipitation chemistry. A valid data base is a prerequisite for a correct interpretation.
- It is impossible to establish in precipitation whether the available NH_4^+, H^+ or Ca^{2+}-ions have originated from HNO_3, H_2SO_4, nitrates or sulfates.[1])
- Backtrajectories calculated for precipitation events cannot always be used for the investigation of separate source regions, especially not if they cover more than one source region.
Measuring the wash-out of the plume of a single source offers some problems too:
- Validation of the data base is problematic.
- Distinction between the contribution of the plume and of the background is very difficult.

Nonetheless these problems must be solved if one desires to estimate the contribution of single sources to the acidification of precipitation.
The background contribution has been investigated by means of a series of experiments performed by ECN and IMOU near Lelystad in The Netherlands.
Special plume wash-out studies are carried out under the plume of the Flevocentrale situated near Lelystad. These studies are a joint enterprise

of KEMA, ECN and IMOU.

Background studies

To tackle the problems mentioned in the introduction we decided to sample precipitation with groups of 8 or more identical rainsamplers, placed in one spot. [2]) In the last experiment the sampling period was 24 hours. Longer periods are unsuitable because the meteorological parameters are varying too much. The samples were analysed for "bulkelements" (Cl^-, SO_4^{2-}, NO_3^-, H^+, NH_4^+, Na^+, K^+, Mg^{2+}, Ca^{2+}) and some trace elements (F^-, Pb, Zn, Cd, Cu).

Sampling with groups of samplers has of course the drawback that a lot of samples have to be analysed but there are advantages:

- Contamination contributed by the samplers, sample pretreatment or analytical procedures can be traced. The standard deviation calculated from the results of a number of identical samplers stays constant or goes down at increasing mean concentration of a given compound. In case of contamination of the standard deviation increases. Comparing figures 1 and 2 makes it clear that a copper contamination is present. This fenomenon has enabled us to test our samplers, leading to necessary new developments (presently we are building the "Mark VI", completely made of polyethylene, with a very tight fitting cover of special design and a specially treated funnel).
  Sample pretreatment was another source. All sample pretreatment now is carried out in clean benches. These measures had a drastic effect: the mean zinc concentration has gone down from about 100 to less than 25 ppb during 4 years of experiments.
  In the same way deficiences in our analytical methodology have been traced. Problems were encountered in the determination of SO_4^{2-}, NO_3^- and NH_4^+ which were solved by using techniques like ion-chromatography and flow-injection analysis. [3,4])
- As 8 or more results are available per sampling period, tests for the elimination of outliers can be applied. This is important because outliers can influence the averages rather drastically, and they deteriorate severely the results of chemometrical techniques such as cluster analysis or multiple regression analysis, both of which are often very useful tools as will be described later.

We decided to use an indirect approach to investigate the role of strong acids and their source regions.
First we tried to divide the sampling periods in different groups, by means of cluster analysis.
Cluster analysis enables the comparison of rain periods by taking the concentration of more than two compounds into account. The principle of this method is that periods with n parameters are ordered in a n dimensional space assigning an n dimensional point per period and constructing groups of periods according to a chosen criterium. The best results were obtained in our case with the so-called centroid method. [5])
Clustering with the concentrations of 12 or 9 compounds per sampling period had no effect on the grouping. The result is given in figure 3 as a so-called dendrogram; the distance between the horizontal line connecting two periods or two groups of periods and the x axis is representative for the differences found between the periods or the groups of periods.

Of the 4 groups, group 1, 2 and 3 are of continental origin and group 4 has a maritime character and this division corresponds nicely with meteorological analysis.
The mean concentrations per group of H^+, NH_4^+, NO_3^-, SO_4^{2-} and Na^+

differ considerably (see figure 4). The standard deviation within each group is typically 10 to 20 % , so the differences are statistically significant. The differences between the groups in sulfate concentration (52 to 78 µmoles l^{-1}) and nitrate (36 to 84 µmoles l^{-1}) are less important as encountered in the cases of H^+ (20 to 120 µmoles l^{-1}), ammonium (36 to 132 µmoles l^{-1}) or sodium (29 to 375 µmoles l^{-1}).

The role of NH_4^+ as neutralizer emerges very clear. The rains of group 3 are very acid indeed (pH 3.9). In figure 5 the relative contribution of each group to the deposition of H^+, NH_4^+, NO_3^- and SO_4^{2-} is given. The group 3 rains contribute nearly 60% of the H^+ deposition, but only 30% of the total rainfall.

The contribution of HNO_3 and H_2SO_4 to the acidity of rain has been investigated by multiple regression analysis in combination with a superimposed model:

$$2x\ [SO_4^{2-}] = A_1.[H^+] + A_2.[NH_4^+] + A_3.2.[Ca^{2+}] + A_4.2.[Mg^{2+}] + \ldots\ldots$$

$$[NO_3^-] = B_1.[H^+] + B_2.[NH_4^+] + B_3.2.[Ca^{2+}] + B_4.2\ [Mg^{2+}] + \ldots\ldots$$

$$[Cl^-] = C_1.[H^+] + C_2.[NH_4^+] + C_3.2.[Ca^{2+}] + C_4.2\ [Mg^{2+}] + \ldots\ldots$$

Superimposed model:
$A_1 + B_1 + C_1 = 1$
$A_2 + B_2 + C_2 = 1$
$A_3 + B_3 + C_3 = 1$
  etc.

The results are given in figure 6, the first bar represents twice the sulfate concentration, the second the nitrate concentration, both in 10^{-6} M. The different shades represent the relative contribution of H^+, NH_4^+ and Ca^{2+} as counter ions.

It is clear that HNO_3 and H_2SO_4 are partly neutralized by ammonia and calcium compounds. Magnesium does not play an important role (the contribution of Mg^{2+}, Na^+, K^+ is given as the unshaded part of the bar of which Mg is by far the most important) with the exception of group 4, the maritime rains. This is not surprising taking in mind that most of the magnesium in precipitation in The Netherlands originates from the ocean. Sodium is for more than 95% assigned to Cl^- by regression analysis, no HCl is detected at all.

The fraction of sulfuric acid and nitric acid compared to the total amount of sulfate and nitrate available differs considerably between the groups (table 1). The last column gives the ratio between NO_3^- and 2 x the SO_4^{2-} concentration.

Table 1   Fraction HNO_3 resp. H_2SO_4 of total NO_3^- and SO_4^{2-} in %.

	% HNO_3	% H_2SO_4	NO_3^-/2x SO_4^{2-}
Group 1	21	21	0.48
Group 2	17	11	0.35
Group 3	41	55	0.57
All continental rains (1+2+3)	36	35	0.50
Maritime rains (Group 4)	47	25	0.30
All rains	38	32	0.44

Remarkable is the relative abundance of HNO_3 in the group of maritime rains and of sulfuric acid in group 3.

The backtrajectories of 41 sampling periods are given in figure 7, with different symbols for each group. There is a clear overlap in the direction of the trajectories of each group. Some tendencies can be distinguished: The trajectories of group 1 are generally of a more southern origin compared to the group 2 trajectories. The main difference is the mean length of the 24 hour backtrajectories per group. The trajectories of the group 4 rains (maritime) are very long, the trajectories of the group 3 rains are the shortest ones, with the trajectories of group 1 and 2 in between. Comparing this situation with the chemical composition per group could lead to the following interpretation:
- The airmasses of the precipitation events belonging to group 4 have moved very quickly over the industrial areas, and correspondingly contain a low load of precursors. As the time for reactions is limited, HNO_3 is predominant because of the high conversion rate of NO_x. Little NH_3 is taken up.
- In case of the group 3 rains more time has been available so a considerable amount of sulfuric acid has been formed which is not completely neutralized by ammonia.
- The group 1 and 2 rains are an intermediate case, neutralization has however proceeded further compared to the group 3 rains, for which we are not able to give an explanation.
Further interpretation is needed, but it seems that the described experimental strategy can lead to a better prediction of source-recepter relation.

Plume-washout experiments
As already indicated in the introduction, validation of the data-base and making a distinction between the specific contribution of the plume and the background to washout of gases and aerosols by precipitation are the main problems in experimental studies of plume-washout.
Hales has described an elegant method to measure the washout of SO_2 in which the washed out sulfite is conserved for analysis by the addition of mercury tetrachloride. [6] The background concentration of SO_2 is low compared to the SO_2 concentration in the plume, so the background contribution is negligible.
This method fails, however, if the washout of other components must be measured or if the primary sulfate concentration in the plume is relatively high.

We have tested a different approach for the "Flevo" power plant near Lelystad. The general set up of the experiment is given in figure 8. Two arcs, consisting of 30 samplers each were placed at a distance of 5 resp. 15 km around the oil fired Flevo power station near Lelystad. The arcs were so situated that the plume would pass the rain samplers if the wind would come from westerly to northeastern directions. Generally maritime air with low concentrations of sulfate, nitrate and precursors can be expected under these circumstances.

The samplers were placed at the sample locations just before the onset of precipitation and removed directly after the end of it, to minimize dry deposition.

During the experiment SF_6 was emitted from one of the stacks of the powerplant. A mobile laboratory traced the plume by monitoring the SF_6 and SO_2 concentration on ground level. If the plume was detected precipitation samples were taken right under the plume by mobile rain samplers. Some of the samples were treated with formaldehyde to preserve sulfite and nitrite. Others were used to measure the H_2O_2 concentration in the precipitation.

All "stationary samples" were analysed for bulkelements and some trace elements (F⁻, Pb, Cd, Cu and Zn).

The variations in the concentration and deposition of bulk elements found in the samples of the stationary samplers were often so large (a decade or more) that it was very difficult to recognize the contribution of the plume.

We tested the following hypothesis. Assuming that a well mixed background air mass is present, the variation in meteorological parameters such as rainfall rate, droplet spectrum etc. should be the most important cause for the variation in concentration. Especially if internal mixed aerosol is available we could expect that all compounds are washed out with the same efficiency, with the result that good correlations must be present between the concentration or the deposition of all background components. This was indeed observed during all experiments, including a high correlation for sulfate with other compounds.

For each sampler a sulfate concentration can be calculated based on the correlation with other species

$$(SO_4(i))_{calc} = A.C(i) + B$$

(C(i) is the concentration of a species not available in the plume).

The mean of these calculated sulfate concentrations is then computed taking into account the "goodness of fit" by means of a weight factor w_i.

$$(SO_4)_{calc} = \frac{\Sigma \ w_i \ (SO_4(i))_{calc}}{\Sigma \ w_i}$$

For background samples $\frac{(SO_4)_{measured}}{(SO_4)_{calculated}}$ varied between 0.9 and 1.1, indicating that:

- the data base probably does not contain many outliers, which is an indirect way to validate the data base.
- the contribution of the background to sulfate deposition under a plume can be calculated.

In experiments where the ratio $(SO_4)_{measured}$ / $(SO_4)_{calculated}$ exceeded 1.3 in some samples it was assumed that the plume had passed the arcs of rain samplers. These samples were deleted for background calculation and the procedure was iterated.

In figures 9, 10 and 11 the observed deposition for sulfate, magnesium and nitrate are given.

Magnesium is present in the plume because magnesiumoxide is added to the oil as a neutralizer for SO_3 and H_2SO_4 for corrosion prevention. In the upperhalf of the figures the observed and calculated depositions are given. The samplers on the first arc are represented by the sample numbers 1 to 30, the ones on the second arc are denoted by the numbers 31 to 60. The calculated ratio between $(SO_4)_{measured}$ and $(SO_4)_{calculated}$ is given in the underpart of the figures.

The extra deposition of sulfate and magnesium in the same samplers on both arcs is evident. This result corresponds very well with the wind direction during the experiment. No washout of nitrate from the plume was observed. The contribution of the plume was maximal 15 μmoles of SO_4 and 16 μmoles of Mg on top of the observed background contribution. This squares nicely with the results of the mobile samplers; we observed a mean concentration of 1 ppm SO_3^{2-} corresponding to 14 μmoles of SO_4^{2-}.

We have made an estimate of the sulfate deposition at 5 and 15 kilometers distance of the powerplant.

The plume width as detected by precipitation sampling on the first arc was 1500 meter and on the second arc 5000 meter. The deposition at 5 km was

on a 1500 x 1 m^2 cross section 2 g $SO_4{}^{2-}$ per hour and at 15 km on a 5000 x 1 m^2 cross section 3.5 g per hour.
The estimate deposition under the plume over 20 km on basis of these data is 50 kg SO_2 per hour or 0.7 % of the SO_2 emission of the powerplant.

The magnesium deposition, calculated in the same way is 6.5 kg per hour or 50% of the estimated magnesium emission.

These results are in accordance with other investigations: The washout of SO_2 is limited, but aerosols are washed out quite effectively.

More experiments are needed to furnish a data base which allows definite conclusions. The experiments will be continued in 1982 near the Flevo powerplant and in 1983/84 near a coal-fired power station.

ACKNOWLEDGEMENT

We are indebted to the Management and the personnel of the Flevo Centrale. Especially the assistance of Mr. H. Klink and co-workers was of prime importance for our plume studies. Dr. J. Reiff and Mr. A. van der Hoek of the Royal Netherlands Meteorological Institute are gratefully acknowledged for the calculation and drawing of the trajectories.

REFERENCES
1. S. Beilke, Acid Deposition, Draft Paper, prepared for "Workshop on Acid Deposition". September 1982, Berlin.
2. W.A.H. Asman, J. Slanina, and J.H. Baard. Water, Air, Soil Poll. (1981) 16, 159.
3. J. Slanina, F. Bakker, A. Bruyn-Hess, and J.J. Möls. Anal. Chim. Acta 113 (1980), 331.
4. J. Slanina, F.P. Bakker, P.A.C. Jongejan, L. van Lamoen, and J.J. Möls. Anal. Chim. Acta 130 (1981), 1.
5. J. Slanina, J.H. Baard, W. Zijp, and W.A.H. Asman. Water, Air, Soil Poll., in preparation.
6. J.M. Hales et al. NTIS Report PB 203 129 (1971).

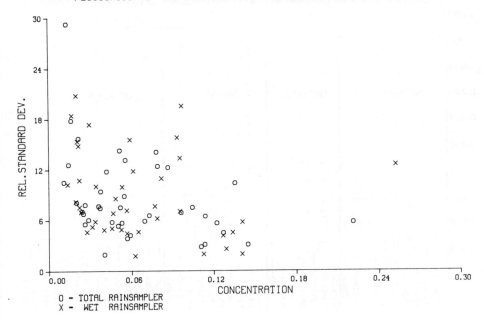

Figure 1

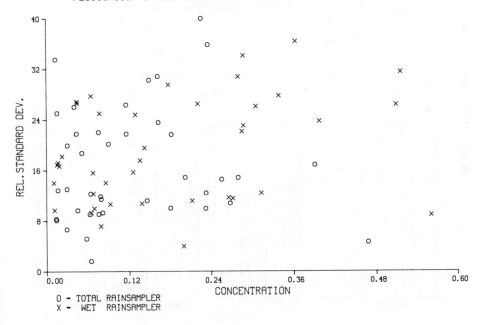

Figure 2

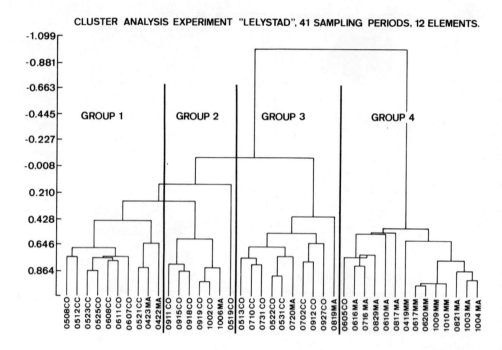

Figure 3

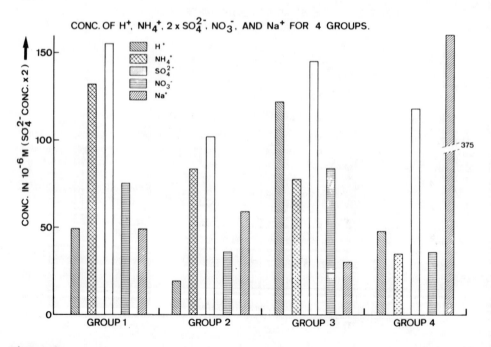

Figure 4

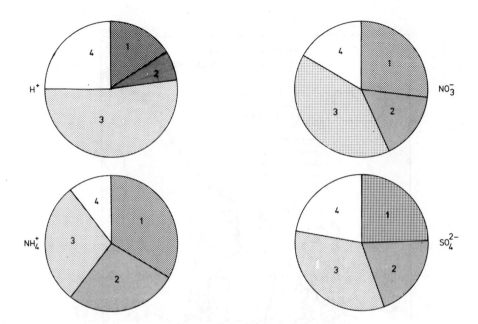

DEPOSITION OF BULK ELEMENTS PER GROUP.

Figure 5

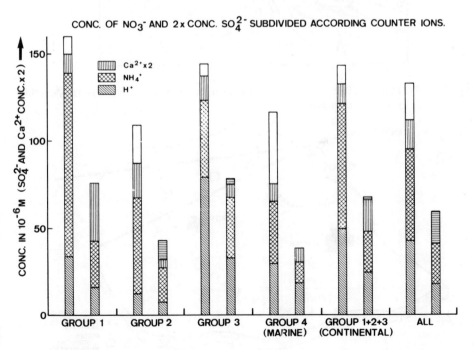

Figure 6

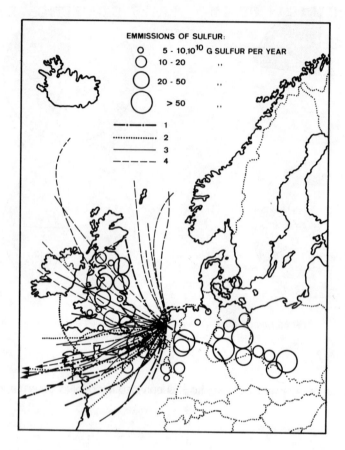

Figure 7 — Trajectories (24 hours) for groups of sampling periods obtained
by cluster analysis

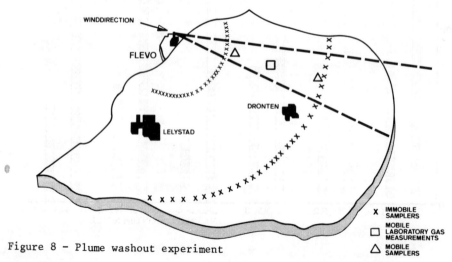

Figure 8 — Plume washout experiment

COMPARISON OF TWO SERIES DEPOSITION DATA

Rainspell number 4951

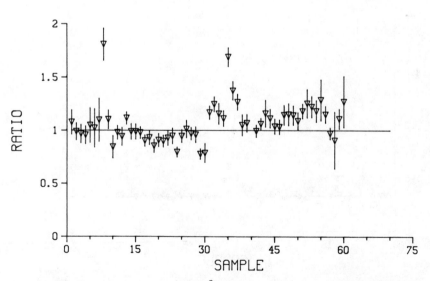

Fig. 9   Deposition of SO_4^{2-} in 60 rainsamplers and ratio of
measured and calculated SO_4^{2-} deposition.

COMPARISON OF TWO SERIES DEPOSITION DATA

Rainspell number 4951

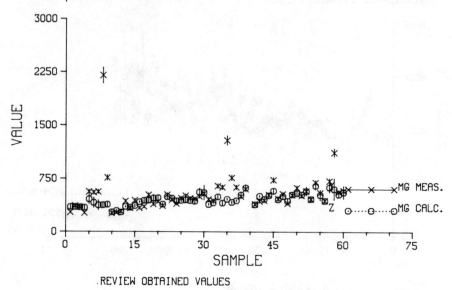

.REVIEW OBTAINED VALUES

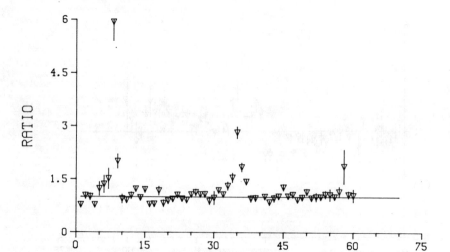

Fig. 10   Deposition of Mg in 60 rainsamplers and ratio of
measured and calculated Mg-deposition.

COMPARISON OF TWO SERIES DEPOSITION DATA

Rainspell number 4951

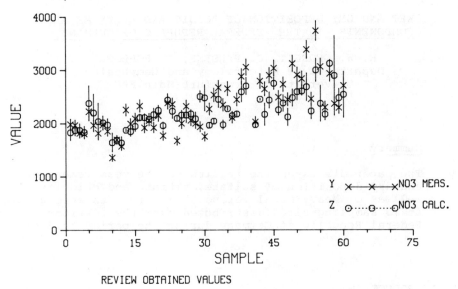

REVIEW OBTAINED VALUES

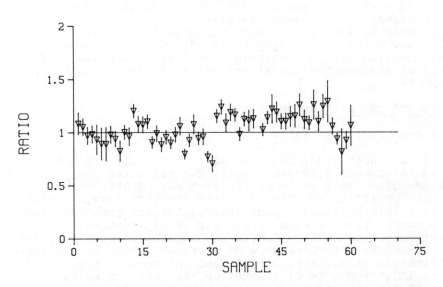

Fig. 11   Deposition of NO_3^- in 60 rainwamplers and ratio of
measured and calculated NO_3^--deposition.

WET AND DRY DEPOSITION OF ACIDIC AND HEAVY METAL COMPONENTS IN THE FEDERAL REPUBLIC OF GERMANY

H.-W. GEORGII, C. PERSEKE, E. ROHBOCK
Department of Meteorology and Geophysics
University of Frankfurt/Main FRG

Summary

The paper discusses the results of the measurement of wet and dry deposition of sulfate, nitrate and chloride as well as of heavy-metal aerosols. The results were gained at 13 sampling-sites distributed over the territory of the Federal Republic of Germany during the period 1979 to 1981.

1. INTRODUCTION

In the scope of the German deposition network, comprising 13 stations of different air quality (GEORGII et al. 1982), dry and wet deposition of various atmospheric trace substances were measured by means of an automatic wet/dry deposition collector during the period 1979-1981. A detailed description of the deposition collector is given elsewhere (GEORGII et al., 1982; ROHBOCK, 1982). Rain was collected in a polyethylene funnel and bottle on a daily basis; dry deposition was sampled in high walled glass containers according to the standard method of BERGERHOFF (VDI-Richtlinie 2119, 1972) during periods of two weeks.

2. RESULTS AND DISCUSSION

Ion-Concentration of Rain:

In a recent study the composition of "acid" rain in the FRG was discussed (PERSEKE, 1982). The rainwater analyses show that pH-values (50% value of the frequency distribution) between $<$ pH 4.2 in a polluted region (Frankfurt/Main) and $<$ pH 4.6 in a less polluted region (Hof/Saale) are observed. The ionic composition of rain is similar at all stations -with the exception of the station Schleswig near the coast of the Baltic sea, where the chlorides show an elevated level. The relative contribution of sulfate, nitrate and chloride to the sum of the anions ($SO_4^= + NO_3^- + Cl^-$) makes up 55 - 60% for sulfate, 25 - 30% for nitrate and less than 15% for chloride on an equivalent basis. These values agree quite well with measurements in the United States (WILSON et al., 1980).

Wet deposition of Acidic Components:

In context with observed damages in forest-ecosystems wet deposition has gained growing attention. The amount of wet deposition is the product of both ion concentration in rain and precipitation amount. Most data indicate that wet deposition rates are proportional to the precipitation rates. On the other hand the concentrations of various ions are often highest with light rain. Low ion concentrations can yield high deposition rates if the precipitation amount is high.

Based on the concentration data and precipitation rates from the different stations of the network the wet sulfur deposition pattern is plotted in Fig.1. It can be seen that highest deposition rates are found during the summer season due to high precipitation amounts. A maximum of the sulfur deposition pattern with values between 3.3 mg S $m^{-2}d^{-1}$ (autumn)- 9.2 mg S $m^{-2}d^{-1}$ (summer) occurs in the Ruhr area, where the highest sulfate concentration values were observed. However, in areas which are considered to be rather unpolluted (like Deuselbach) the wet sulfate deposition still amounts to 50-60% of the wet deposition in the Ruhr area. Particularly, the mountain stations receive high rates of wet deposition. High precipitation rates caused by orographic rainfalls determine the wet deposition in mountain regions.

An evaluation of the wet deposition with respect to wind direction shows that the highest percentage of the yearly deposition is found with advection from southwest to west, as westerly winds are prevailing during frontal precipitation.

For the wet nitrate deposition a similar pattern with maximum values during summer (3.0 mg N $m^{-2}d^{-1}$ in the Ruhr area, 3.6 mg N $m^{-2}d^{-1}$ in Hamburg) is observed. Lowest nitrate depositions are found in the less polluted regions of Hof and Deuselbach.

Dry deposition of Acidic Components:

In addition to wet deposition dry deposition is a main sink of atmospheric trace gases and aerosoles. The dry deposition of particles depends on the particle size (CLOUGH, 1973). Small particles < 0.1 µm are removed by Brownian motion, whereas particles greater than 1 µm are deposited due to sedimentation. Particles in the size range 0.1-1 µm are less efficiently removed by dry deposition. These particles are mainly involved in precipitation processes acting as cloud-condensationnuclei.

Dry deposition of particles (dustfall) was sampled on an artificial surface in the BERGERHOFF glass-container.

In Fig.2 a comparison of wet an dry sulfate deposition is presented. It can be seen that the dry sulfate deposition accounts only to 9-29% of the total sulfate deposition. Dry deposition rates of particulate sulfate up to 1.9 mg S $m^{-2}d^{-1}$ are measured in Essen.

As can be seen from Fig.3 the wet deposition is the dominant deposition mechanism for nitrate, too. The dry nitrate deposition with 8-23% of the total nitrate deposition is less

Wet Sulfate - Sulfur Deposition

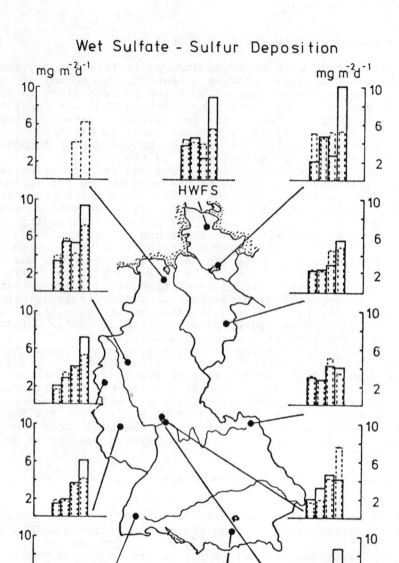

Fig. 1 Regional and seasonal distribution of the wet
sulfate-sulfur deposition in Germany during the
period 1979-1981

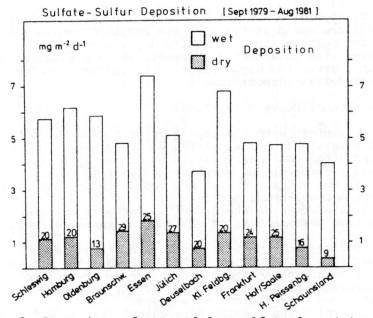

Fig. 2   Comparison of wet and dry sulfate deposition
(figures: % - dry deposition)

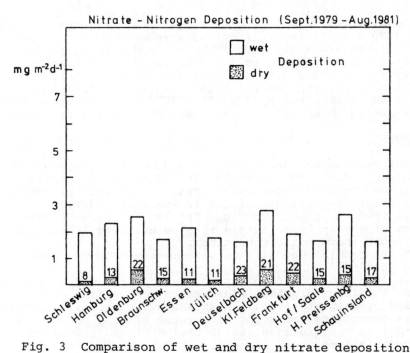

Fig. 3   Comparison of wet and dry nitrate deposition
(figures: % - dry deposition)

efficient.

It should be mentioned, that measurements of dry deposition of sulfates and nitrates with the BERGERHOFF-method account only for particulate matter. The dry deposition of gases is not measured with this method. However, it is estimated that in polluted areas the dry deposition of sulfur-dioxides exceeds the wet sulfate-deposition.

Deposition of Heavy Metal-Ions:

In context with the impact of acid deposition on terrestrial ecosystems the deposition of trace metals has gained growing attention. Heavy metals exist mainly in the form of aerosol particles; the gasphase is less important.

Dry and wet deposition are two dominant removal processes for the aerosol-bound metals.

In the scope of the deposition network dry and wet deposition measurements of various substances were also carried out.

The dry deposition of aerosol particles depends largely on the particle size (CLOUGH, 1973, GARLAND, 1982). Therefore the evaluation of the relative importance of dry and wet deposition was concentrated on metals (Pb, Cd, Mn, Fe), which are bound on particles of different size range.

The following results were gained for the different metal-ions:

LEAD:

Lead, which is mainly observed in the submicron particle range (MMD 0.4 μ), is preferentially removed by wet deposition. In unpolluted areas dry deposition of lead amounts to 6-8% of the total deposition. In polluted areas (Frankfurt/Main) the dry deposition rates are increased (up to 35% of total deposition). For the Federal Republic average total lead deposition rates between 40 μg Pb m^{-2}d^{-1} in unpolluted areas and 87-100 μg Pb m^{-2} d^{-1} in the Rhein Main area were found.

CADMIUM:

For cadmium wet deposition is a dominant sink, too. Dry deposition accounts for less than 20% of the total cadmium deposition. Only in urban areas (Frankfurt/Main) the contribution of dry cadmium deposition is increased (30%). This can be explained by gaseous cadmium. Cadmium as the most volatile species of the analyzed metals is also existing in the gasphase under atmospheric conditions.

Total cadmium depositions were measured in the range 1-4 μg Cd m^{-2}d^{-1}.

As lead and cadmium are predominantly removed by precipitation, the input of these elements to ecosystems can be assessed by rainwater analysis (NÜRNBERG et al., 1982). These authors determined wet deposition rates of lead between 28.8 μg Pb m^{-2}d^{-1} in unpolluted areas (Deuselbach) and 116 μg Pb m^{-2}d^{-1} in industrialized areas (Essen).

MANGANESE:

A different deposition pattern is found for manganese and iron, which occur mainly in the coarse particle mode. These particles are removed more effectively by dry deposition. For manganese-containing particles both removal processes are of similar importance. This result is in agreement with measurements by PATTENDEN et al.,(1982). Only at mountain stations which receive increased precipitation rates, the wet deposition dominates. At the stations of the German deposition network total manganese deposition rates between 30-50 µg Mn $m^{-2}d^{-1}$ are found. Higher deposition rates (143 µg Mn $m^{-2}d^{-1}$) are measured in the vicinity of iron processing industry.

IRON:

Iron-containing particles are predominantly coarse. The majority (>70%) is removed by dry deposition. The total iron-deposition varies between 200-500 µg Fe $m^{-2}d^{-1}$ in unpolluted areas and 800-1500 µg Fe $m^{-2}d^{-1}$ in urban and industrialized areas.

3. CONCLUSIONS

Mass balances show the different importance of dry and wet deposition for atmospheric metals. Metals bound on submicron particles (Pb, Cd) are primarily removed by wet deposition, whereas metals bound on coarse particles (Mn, Fe) are mainly removed by dry deposition.
The accumulation of manganese and iron-particles is restricted to the vicinity of the source.
In contrast to that, the deposition pattern of the highly toxic lead and cadmium particles follows the precipitation pattern. With regard to the ecological impact of lead and cadmium their high solubility has to be taken into consideration.

Acknowledgement

The research reported in this paper has been sponsored by the Federal Environmental Agency (Umweltbundesamt) under contract 104 02 600 which is acknowledged with thanks.

4. LITERATURE

CLOUGH, W.S., (1973)
   Transport of particles to surfaces,
   Aerosol Science 4, 227-234

GARLAND, J.A., (1982)
   Field Measurements of the Dry Deposition of Small
   Particles to Grass
   H.W. GEORGII, J. PANKRATH (Ed.) Deposition of Atmos-
   pheric Pollutants, D.Reidel Publ. Comp. 9-17

GEORGII, H.-W., PERSEKE, C., ROHBOCK, E., (1982)
   Feststellung der Deposition von sauren und langzeitwirk-
   samen Luftverunreinigungen aus Belastungsgebieten,
   Abschlußbericht im Auftrag des Umweltbundesamtes
   (Forschungsprojekt 104 02 600)

NÜRNBERG, H.W., VALENTA, P., NGUYEN, V.D., (1982)
   Wet Deposition of Toxic Metals from the Atmospheric
   Pollutants, D. Reidel Publ. Comp. 143-159

PATTENDEN, N.J., BRANSON, J.R., FISHER, E.M.R., (1982)
   Trace Element Measurements in Wet and Dry Deposition
   and Airborne Particulate at an Urban Site
   H.-W. GEORGII, J. PANKRATH (Ed.) Deposition of Atmospheric
   Pollutants, D.Reidel Publ. Comp. 173-187

PERSEKE, C., (1982)
   Composition of Acid Rain in the Federal Republic of
   Germany -Spatial and Temporal Variations during the
   Period 1979-1981
   H.-W. GEORGII, J. PANKRATH (Ed.) Deposition of Atmos-
   pheric Pollutants, D. Reidel Publ. Comp. 77-87

ROHBOCK, E., (1982)
   Atmospheric Removal of Airborne Metals by Wet and Dry
   Deposition
   H.-W. GEORGII, J. PANKRATH (Ed.) Deposition of Atmospheric
   Pollutants, D. Reidel Publ. Comp. 159-173

WILSON, J., MOHNEN, J., (1980)
   Wet Deposition in the Northeastern United States, ASRC
   Publication 796, State University of NY, Albany

HEAVY ELEMENTS IN ACID RAIN

C. RONNEAU and J-Ph. HALLET
Laboratoire de Chimie Inorganique et Nucléaire
Université Catholique de Louvain
Belgium

SUMMARY

Acidity in rain is the indicator of a more géneral state of pollution : indeed, heavy elements (H.E.) are deposited on soils together with the anions responsible for rain acidity. The effects of acid rain on the environment begin to be fairly well documented [1,2,3]. H.E. are probably more controversial as evidence has still to be provided as to their eventual adverse effect on man and the environment, at least in remote rural regions of industrialized countries [7].

1. INTRODUCTION

Three years ago, rain gauges were installed in various sites throughout Belgium, in order to determine the chemistry of deposition in various environmental conditions. Some gauges were installed in reputedly "clean" zones of the country, mainly in the Ardennes (figure 1) while others were placed downwind from the industrial zone of Charleroi (figure 2). The latter gauges were aligned following a north-eastern axis under prevailing dry and rainy winds from industrial complexes, mainly metallurgical plants with coke production, blast furnaces, steel making, foundries,... as well as electricity production and, of course, domestic heating [4].

Whenever possible, the gauges were installed in farms in order to allow the establishment of an elemental deposition budget on the pastures and to allow the observation of eventual relations between deposition and transfer towards the human food chain (dairy products, meat,...). All these gauges collected total deposited material, i.e., dry plus wet deposition.

Rain water was analyzed for different contaminants :
- acidity by a glass electrode,
- anions by ion chromatography (DIONEX),
- Cd and Pb by atomic absorption spectrometry,
- 18 other elements by instrumental neutron activation analysis.

2. RESULTS AND DISCUSSION

Geographical profiles
Figure 3 shows the pH profile observed along the gauges axis : pH is at its maximum in the center of the industrial zone. In fact, lime and ammonia, which are emitted by the iron industry efficiently, neutralize the acidity of primary gases such as SO_2, NO_x and HCl [5].

Many metals follow the same profile (see figure 4), i.e., their maximum deposition is observed at a short distance from the industrial center and the profile rapidly decreases to attain, after about 20 km, the deposition rates ($\sim$ 20% of the maximum) observed in the purely rural, remote regions of the country (the Ardennes). It is interesting to note that these metals are relatively not volatile.

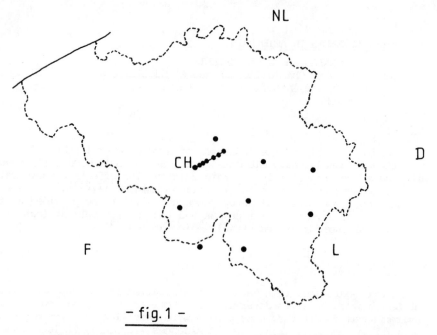

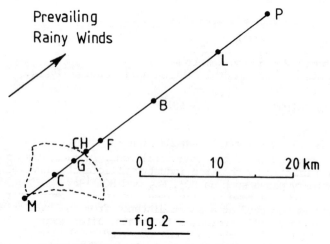

NL

D

CH

F

L

— fig. 1 —

Localization of the rain gauges over Belgium (rural areas,
except in the vicinity of CH : Charleroi)

Prevailing
Rainy Winds

P

L

B

CH F

G

C

M

0   10   20 km

— fig. 2 —

Localization of the gauges along a N-E axis from the industrial
zone of Charleroi (------ limits of the zone)

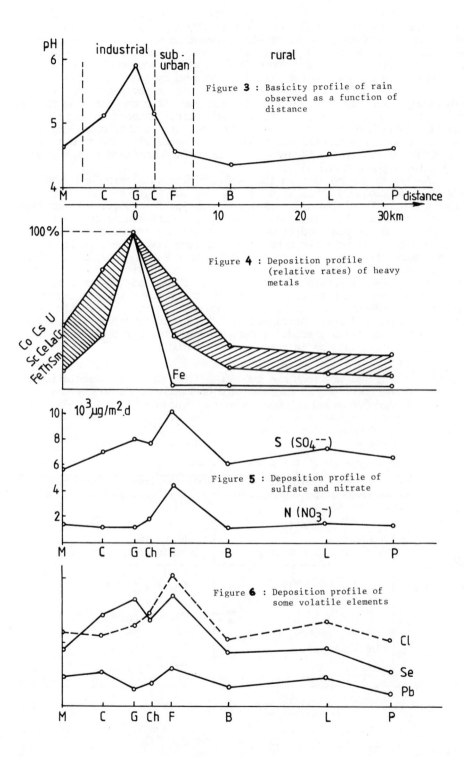

Figure **3** : Basicity profile of rain observed as a function of distance

Figure **4** : Deposition profile (relative rates) of heavy metals

Figure **5** : Deposition profile of sulfate and nitrate

Figure **6** : Deposition profile of some volatile elements

On the other hand, acid gases such as SO_2 and NO_x are the main precursors of acidity in rain; it was interesting to trace the deposition behaviour of anions such as sulfate and nitrate which under the form of sulfuric and nitric acids are recognized to be the source of protons in rain water. Figure 5 shows the profile displayed by these anions : a profile somewhat different from the previous one in the sense that the maximum deposition is observed 5 km further away than was the case for pH. Another characteristic is that profiles are more flattened as compared with the profiles of pH and non volatile elements. In this case, deposition in the Ardennes amounts to 20-60% of the deposition observed in the very center of Charleroi. Clearly, in this case, deposition is far from being a local phenomenon! [6].

Finally, as can be seen in figure 6, chlorine, selenium and lead show the same tendency. This is also true for Cd, Na, F and Zn. Other elements behave even more diffusely, for example, Ba, Sb,... Interestingly, these elements are characterized, either by the diffuseness of their sources (Pb, Ba, as fuel additives) or by their higher volatility which could lead these elements to behave more or less like gases, at least during a part of their existence in the atmosphere. Condensation of these elements under the form of very thin particles could similary explain the parallelism observed between their deposition rates and the rates of gas-derived anions such as SO_4^{--} and NO_3^-. It should be interesting to check if, at greater distances of pollution sources, these volatile elements are more closely correlated with rain acidity which, once again, depends mainly on primary acidic gases, provided basic materials are not added to the pollutants as is the case of Belgium.

Correlations in rural zones

As far as rural stations are concerned, the elemental associations observed in the geographical profiles are partially confirmed by the intercorrelations observed between elemental concentrations in individual samples. On can distinguish two groups of intercorrelated elements :
- non volatile elements like Fe, Sc, Cr,...
- volatile metals and anions such as SO_4^{--}, NO_3^-, Cl^-, Se,...

Table I gives a summary of these associations.

Table I : Associations of elements with anions and iron in deposition as observed along a geographical axis (A) and in time, in a rural station (B).

with	CORRELATIONS	
	A) in space (downwind from a city)	B) in time (samples from a rural station)
NO_3^- SO_4^{--}	Br, Cl^-, F^-, Na, Se Diff, Ba, K, Pb, Sb, Zn	Br, F^-, K, Sb
		Ba, Ce, Co, Cs, Se, Sm
Fe	As, Ce, Co, Cr, La, Sm, Th, U , Cs	Cr, Sc, Pb
	Basicity	Basicity ?

- 152 -

3. CONCLUSIONS

Acidity in rain can be effectively neutralized by anthropogenic or natural bases such as ammonia and lime derivatives. Toxic elements, on the contrary, are likely to retain their harmful potential all along their dispersion paths in the environment. Except for changes in solubility due to minor chemical evolution, dilution is the only process to really alleviate the hazard they impose to the biosphere. As a matter of fact, dilution is performed at the expense of very large areas of agricultural soils, as it was demonstrated that even remonte rural zones of Belgium are submitted to non negligible deposition fluxes of the most toxic among the H.E. As these toxic elements are, by chance, the most volatile, one can expect that they will be spread to the some extent as that of sulfates and nitrates.

Heavy elements in acid rain should be thoroughly considered, even thoroughly as acidity itself. To emphasize this remark, figure 7 presents the relationship observed in eight farms of Belgium between the deposition of selenium on pastures and the selenium content of milk from cows grazing on these pastures. Clearly, elemental deposition on European soils is not solely a sink of air pollutants but, rather, the starting point of concern for toxicologists.

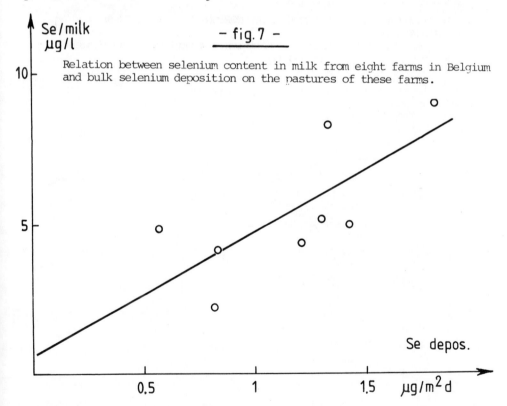

Se/milk µg/l

– fig. 7 –

Relation between selenium content in milk from eight farms in Belgium and bulk selenium deposition on the pastures of these farms.

Se depos.

µg/m²d

4. REFERENCES

(1) L.S. EVANS, G.R. HENDREY, G.J. STENSLAND, D.W. JOHNSON and A.J.FRANCIS
Acidic precipitation : considerations for an air quality standard.
WASP, 16, (1981) 469-509

(2) N.R. GLASS, G.E. GLASS and P.J. RENNIE
Effects of acid precipitation.
Envir.Sci.Techn., 13 (1979) 1350-1355

(3) G.E. LIKENS, R.F. WRIGHT, J.N. GALLOWAY and T.J. BUTLER
Acid rain.
Sci.Amer., 241 (1979) 39-47

(4) J-Ph. HALLET, P. LARDINOIS, C. RONNEAU and J. CARA
Elemental deposition as a function of distance from an industrial zone.
To be published in Sci.of the Total Envir. in 1982

(5) Ch.D. HENDRY and P.L. BREZONIK
Chemistry of precipitation at Gainesville, Florida.
Envir.Sci.Techn., 14 (1980) 843-849

(6) M.R. MOSS
Spatial patterns of precipitation reaction.
Envir.Pollut. 8 (1975) 301-315

(7) J.N. GALLOWAY, J.D. THORNTON, S.A. NORTON, H.L. VOLCHOK and
R.A.N. Mc LEAN
Trace metals in atmospheric deposition : a review and assessment.
Atm.Envir., 16 (1982) 1677-1700

ACIDIC INPUT TO A BEECH AND SPRUCE FOREST

G. GRAVENHORST
Laboratoire de Glaciologie et Geophysique de l'Environnement
2 Rue Très Cloîtres   F 38031 Grenoble

K.D. HÖFKEN and H.W. GEORGII
Institut für Meteorologie und Geophysik
Feldbergstr. 47   D 6000 Frankfurt a.M.

Summary

In a wooded area in north-west Germany the chemical composition of atmospheric aerosols and rainwater sampled
above and below a beech and a spruce canopy were analysed
to investigate dry and wet deposition fluxes from the
atmosphere to the vegetation-soil system. The concentrations of the aerosol components were by 17% - 41%
smaller below than above the canopy indicating a considerable aerosol uptake by the tree-canopies. This
aerosol uptake showed a minimum for components with a
mass mean diameter (MMD) of about 1 μm for both spruce
and beech trees. The concentrations of rainwater constituents were considerably higher in the throughfall
below the canopy than in the open field. These increases
were most pronounced for the spruce forest. The soil of
the spruce forest received a free acidity input via
throughfall about 5 times as high as it was brought
down in rainwater above the canopy. It could be shown
that most of these increases are due to dry deposition
of gaseous and particulate matter onto the vegetation
during the periods without rain. For aerosol components
a dry deposition velocity for a spruce and beech forest
between 0.4 cm s^{-1} and 2 cm s^{-1} could be determined. A
minimum was found for Mn (MMD of about 1μm) in agreement with the dependence of the filter effect on particle size. It seems that dry deposition of gaseous sulfur dioxide and not wet deposition of acid rain contributes most to the input of free acidity to a spruce
forest ecosystems in this rural area.

1. INTRODUCTION

To understand a possible impact of atmospheric constituents on ecosystems it is necessary to know the pathways via
which the constituents interact with the ecosystem and to
estimate the involved flux rates. Input of acidity by rain
to aquatic and terrestrial ecosystems has gained considerable public and scientific interest during the last years.
Up to now, however, it is not known whether forest ecosystems in Germany receive more acidity by dry deposition of
gaseous and particulate trace substances or by wet deposition

in rain. This uncertainty is due to the fact that the soil-
vegetation-atmosphere interface can hardly be described by a
theoretical model and that the results of flux measurements
made under idealized conditions in the laboratory or in the
field do not comprise the wide spread non-ideal environments
in the field. One possible approach to overcome this problem
is to use the canopy-atmosphere interface itself as a measuring tool thus integrating over all individual large and small
scale processes without sophisticated model assumptions. Such
an approach has been used to study flux rates of airborne
trace constituents to forest ecosystems via wet and dry deposition (Ulrich et al 1976, Ulrich et al 1979, Mayer and
Ulrich 1982). We tried to extend their comprehensive studies
by including chemical analysis of the airborne aerosol above
and below the forest canopy. Furthermore the contribution
of leaching from the leaves to the composition of throughfall
was estimated. From flux differences in rainwater for components without a gaseous precursor in the ambient air the dry
deposition rates and dry deposition velocities for aerosol
components can be derived. Thus the different processes contributing to the throughfall composition at the forest floor
could be assessed. If effects of atmospheric constituents on
trees are caused by processes affecting the forest soil then
the total input of trace substances to the forest soil is of
special concern and not only the flux of trace substances
in rainwater above the forest canopy. Within the framework
of a deposition network in Germany (Georgii et al 1980) this
field study was undertaken. It is described in detail by
Höfken et al (1981).

2. EXPERIMENTAL SET UP

2.1 Field-site description

The forest field stations are situated in the national
preserved park "Solling-Vogler" about 150 km ENE of the Ruhr-
district and about 150 km W of the browncoal and industrial
area between Leipzig and Magdeburg. During the measurements
from Februar to October 1980 the prevailing wind directions
were southwest to west. The site elevation is about 500 m asl
about 30 m lower than the highest elevation of this wooded
resort area 1.5 km to the south. In 1980 the climate station
of the German Weather Service, 2.5 km to the west, reported
65 days with a range of sight below 1000 m often indicating
clouds incumbent on the hills. Only some small villages are
scattered in that area so that the site is free of direct
influence of nearby large pollution sources. Because of the
relatively high elevation in north west Germany the rainfall
rate amounts to nearly $1000 \, l \, m^{-2} \, a^{-1}$ (Matzner 1981). The site
is part of a research area of the Institut für Bodenkunde und
Waldernährung and of the Institut für Geobotanik of the University in Göttingen. It was intensively investigated in the
framework of the International Biological Program and is described in detail by Ellenberg (1971). The two sampling stations in about 500 m distance are situated in a 125 year old
beech forest (Fagus silvatica) and a 90 year old spruce forest
(Picea abies). Two towers extend beyond the 25 to 30 m high

tree stands by about 5 to 10 m. Because of security reasons
it was not always possible to climb on top of the towers. The
open field sampling could, therefore, performed on a wide
meadow about 1.5 km to the west.

2.2 Aerosol measurements

The aerosol was sampled and size fractionated by means
of a 4-stage rectangular impactor. It was designed after
collection criteria given by Marple and Willeke (1975). The
aerosol was impacted on teflon filters mounted on the different stages or collected on a back-up teflon filter. The
beladen filters were stored in polyethylene bottles and kept
just above freezing until dissolved in destilled water and
analysed. The impactor was mounted on a turnable head so that
an attached weather-vane could rotate the intake opening always in the wind direction. The diameter of the intake opening was chosen corresponding to the prevailing wind speed.
As far as possible isokinetic sampling conditions were thus
achieved allowing to compare aerosol measurements made above
and below the canopy under quite different wind conditions.
Each measurement lasted for about 3 - 4 days. The sampling
was interrupted at high relative humidity when fog droplets
bursted on the impactor stages and made interpretations of
the measurements impossible. In supplement the total aerosol
was sampled on membran filters (cellulose nitrate, 0.45 μm
pore size for heavy metals, and teflon, 0.2 μm pore size for
soluble ions). The filter (47 mm diameter) were installed
1.8 m above the ground and used for about one week at a flow
rate of about 1 m^3h^{-1}.

2.3 Rainwater measurements

The rainwater was sampled in polyethylene funnels and
bottles which were manually exposed to the precipitation
shortly before rain started. Below the canopy about 8 samples
were installed randomly to compensate the inhomogeneous distribution of rainfall rate and of chemical composition of
individual throughfall samples. To prevent dissolution of
tree debris in sampled rainwater the sampling bottles were
protected by a fine polyethylene mesh. For heavy metal analysis the rainwater was sucked through membran filters
(cellulose nitrate, 0.45 μm pore size) and filtrate and insoluble residue were analyzed separately. The samples were
kept frozen until analysis to prevent chemical changes during storage (Müller et al 1982). When the rainwater could
not be sampled on the tower just above the canopy the samples
were taken on the open meadow. The rainfall rate reaching the
canopy was, however, allways determined on the meadow because
here our funnel collected 98 ± 3% of the amount registered
by a standard rainfall rate monitor (Lambrecht 1507) whereas
on the tower just above the canopy the collection was generally lower especially at high wind speeds and low rainfall
rates.
The soluble ions F^-, Cl^-, NO_3^-, $SO_4^=$ were analysed by
ionchromatography. NO_3^- and NH_4^+ concentrations were determined
by colorimetric methods (Technicon Autoanalyzer). The sol-

uble and unsoluble fraction of the heavy metals Mn, Fe, Cd and Pb were measured by means of atomic absorption spectrometry (Perkin-Elmer HGA 76).

3. RESULTS

3.1 Aerosol measurements

The concentrations of aerosol components sampled in the open field fall within the range reported for rural areas in Germany (Georgii et al 1981). The aerosol concentrations below the canopies are significantly lower than above (Tab. 1). These smaller concentrations indicate that leaves, needles and branches of the trees filter the air passing through the canopy quite effectively. For the beech stand without foliage a concentration difference of 10% (Mn) to 20% (Cl^-) is found. These differences increase to 25% (Mn) and to 37% (NO_3^-) when the trees are covered with leaves. An even higher filtering effect is caused by spruce trees. The different surface structure of spruce trees compared to beech trees may be responsible for this enhanced aerosol decrease in the spruce forest either by a higher resistance for turbulent air exchange between the air layers above and below the canopy or by a more efficient deposition of particles on spruce needles than on beech leaves.

Distinct aerosol components showed a different reduction in concentration by the canopy (Fig. 1). These differences can be due to the fact that each aerosol component is distributed in its own way over the particle size spectrum and that the dry deposition efficiency depends on the particle size. A first indication whether an aerosol component is bound to smaller or to larger particles is its mass mean diameter (MMD). 50% of the mass of a component is attached to particles with diameters larger than the MMD and 50% with diameters smaller than the MMD. In Tab. 2 the MMD for various components determined between May and October 1980 are listed. The filtering effect avaraged for spruce and beech canopies are shown in Fig. 2 as a function of MMD. From a minimum at about 1 μm (Mn) the concentration difference increases to smaller and larger mass mean diameters. This filter function indicates that particles are deposited on the trees but it does not give the dry deposition flux of aerosols to the canopy of the forests.

3.2 Rainwater measurements

The determined fluxes of trace compounds in rainwater and throughfall are comparable to those reported by Ulrich et al (1979) and Heinrichs and Mayer (1980) for the same site. For Cd, however, we found on the average a concentration lower by about a factor of 6. The rainwater collected in the open field showed similar concentrations as found by Nürnberg et al (1982) and Georgii et al (1982) in rural areas of Germany. The relative contribution of $SO_4^=$, NO_3^- and Cl^- to the sum of these three anions is given in Fig. 3. All concentrations of trace substances were considerably higher in the throughfall than above the canopy (Fig.4a). This increase was stronger pronounced under the spruce than under the beech canopy. The free proton concentrations in the rainwater in the open field sampled in 1980 are comparable to those measured by Ulrich et al (1979) during the years 1968 to 1974 (Fig.4b). They fit into the range of values reported by Georgii et al (1981) and Nürnberg et al (1982) for various regions in Germany. In the throughfall the free proton concentration under the spruce canopy is by a factor of 4.8 ± 1 higher than above the canopy indicating an enormous acidification of rainwater when passing through the canopy. Such an acidification was not observed under the beech trees in the growing season when leaves could interact with the rainwater and the free proton concentration decreased to about 60% of the value above the beech canopy. This decrease was attributed to cation exchange of H^+ and Ca^{++} and Mg^{++} in the leave cells (Ulrich et al 1979). During the months February, March and April without the beech leaves the H^+ concentration increased by a factor of about 1.4 when passing through the beech stand. The soil of the beech forest and the spruce forest neighboring each other receive, therefore, free proton fluxes in rainwater which differ by a factor of about 10 during the growing seasons and of about 3.4 during the winter season. The question whether this large difference in free acidity input to the forest soil is reflected in different soil properties at these two adjacent sites was not investigated in this project.

For Cd, Mn Pb and Fe not only the dissolved but also the insoluble portions were analysed (Tab. 3). The values for the open field are in agreement with measurements reported by Müller (1979) and Nürnberg et al (1982). Below the canopy the insoluble fraction increased for Pb, Cd and Fe. Only the insoluble Mn fraction decreased when the rain passed the canopy. The relative increase of the insoluble fraction of Cd, Pb and Fe in the throughfall was higher than the relative increase in total concentration whereas the insoluble fraction of Mn decreased. This indicates that Mn in throughfall is strongly influenced by leaching from the leaves since only dissolved components should be leached from the leave cells thus increasing the dissolved fraction. Dry deposition of aerosol particles on the vegetation, however, should decrease the dissolved fraction, since components bound to aerosol samples near the earth surface show a much higher soluble fraction than incorporated in rainwater (Müller 1979, Georgii et al 1981).

In a field test to study leaching from the canopy spruce and beech branches, which were protected and unprotected against dry aerosol deposition, were rinsed with natural rainwater (Höfken et al 1981). From the measured concentrations in the washing solution it could be concluded that only for Mn, Cl and NH_4 a leaching effect could be detected. Furthermore the dry deposition of gaseous and particulate matter to a spruce branch seems to be much more efficient than to a beech branch. Because the size and surface area of the spruce and beech branches were different it is only possible to compare the results qualitatively.

4. DISCUSSION

The determined differences in aerosol concentrations above and below the canopies demonstrate that the air is filtered quite efficiently when passing through the canopy. The deposition rate of particles can, however, only be calculated from these data, when the turbulent diffusion resistance between the air layer above and below the canopy is known. In our investigation micrometeorological measurements to determine such a resistance were not made. The aerosol dry deposition fluxes were, however, deduced from the rain chemistry data. Knowing the aerosol dry deposition and the total dry deposition it is possible to estimate the gaseous dry deposition as the difference between these two. The total dry deposition was estimated from the flux difference between throughfall and rainfall above the canopy. For Pb an additional flux could be considered. Analysis of litter fall at the same site for the years 1974 to 1977 showed (Heinrichs and Mayer 1980) that Pb can be irreversibly absorbed by the canopy from the atmosphere since it is quite unlikely that the Pb content in leaves is derived from the soil (Arwik and Zindahl 9174, Ernst 1980). The total dry deposition for lead was correspondingly enlarged by the Pb content of the annual litter fall. The leaching contribution of Mn, Cl$^-$ and NH$_4^+$ to the flux difference in rainwater above and below the canopy were substracted from the throughfall flux according to the relative contribution determined in the leaching field test. Since the relative contribution of leaching to the throughfall concentration depends on the concentration of the rain above the canopy as well as on the length of the dry period between two precipitation events these applied corrections are first estimates only. The resulting total dry deposition rates are given in Tab. 4 and are compared with wet deposition rates. The dry deposition (gaseous and particulate) seems to be more effective for the spruce canopy than for the beech canopy. Especially the high dry deposition for the toxic heavy metals Cd and Pb is striking. The dry deposition rate of gaseous and particulate sulphur-components leading to sulfate in the throughfall is by a factor of about 4 higher than the flux in rainwater above the spruce canopy. Dry deposition of sulfur components is, therefore, a major candidate for acidifing the throughfall of the spruce stand.

To separate particulate from gaseous dry deposition those components are first discussed which can reach the vegetation canopy only in the form of particulate matter because their gasphase concentration can be neglected and concentrations of gaseous precursors are insignificently low at the field site. We assume that Cd, Pb, Mn and Fe belong to this category. Their dry deposition fluxes in Tab. 3 can, therefore, be attributed to particulate dry deposition only. Since the dry deposition velocity v is defined as

$$F = v \cdot c$$

(F: Flux rate of a component with concentration c at a reference height)

The deposition velocities for these components can be determined. They are given in Fig. 5-8 as a function of their MMD. For Mn the dry deposition velocity could be estimated from the insoluble fraction in aerosol and in rainwater samples. Thus the uncertainty introduced by the high leaching of Mn could be avoided. The deposition velocities for those components which can at least be partly produced by gasphase interactions with the leaves and needles are interpolated according to their MMD determined by impactor measurements.

The thus determined deposition velocities are considerably larger than measured and calculated for smooth surfaces (Sehmel 1980, Garland 1982, Marggrander and Flothmann 1982, Davies and Nicholson 1982). Dry deposition velocities seem to increase, however, with increasing roughness height and friction velocity (Sehmel and Hodgson 1974). At our measuring site these two parameters are relatively large compared to field studies over grass surfaces and laboratory measurements. Furthermore the deposition velocity, here determined, does not characterize a small paricle size intervall but a MMD of an aerosol component distributed over a large particle size spectrum. The effectiv interception of fog droplets is also included in our dry deposition flux increasing the dry deposition velocity. The aerosol filter effect (Fig. 2) and the aerosol dry deposition velocity (Fig. 5-8) show a quite similar dependence on MMD although they were determined by independent measurements. This consistency seems to support our measurements.

The gaseous dry deposition should be the difference between the total dry deposition and the aerosol dry deposition. In Fig. 9 the ratio between gaseous dry deposition to particulate dry deposition is given during two measuring periods for the spruce and beech canopy. It seems that the gaseous dry deposition is predominant for the spruce canopy whereas for the beech canopy both deposition mechanismus are likely of similar importance. Taking a SO_2-concentration of about 10 ug m^{-3} in summer and of 20 ug m^{-3} in winter in reference to the rural air monitoring stations Deuselbach and Langenbrügge(Umweltbundesamt, Berlin), and as representative for the Solling area SO_2-deposition velocites of 0.9 cm s^{-1} (summer) and of 2.8 cm s^{-1} (in winter) are deduced for the spruce canopy and of 0.5 cm s^{-1} (summer) and of 0.2 cm s^{-1} (winter) for the beech canopy. These values do not seem to be unrealistic. They do not include the direct dry deposition to the forest floor since the fluxes were only derived from aerosol and rainfall data above and below the canopies. These results should be examined carefully and gasphase measurements above and below the canopy should supplement the analyses of the airborne particulate and liquid phases.

5. CONCLUSION

The results of this investigation characterize two adjacent sites (a spruce and a beech stand) within a widespread forest area of the Solling in north west Germany. Because of its relatively high elevation fog occurs rather frequently and the rainfall rate amounts to roughly 1000 l a^{-1}. The forest canopy seems to be a rather effective sink for aerosol particles. The aerosol concentration was by 10% (Mn) - 41% (Cd) smaller below the canopy than above. This decrease was more pronounced in the spruce forest than in the beech forest. The beech canopy reduces the aerosol concentration twice as much during summer as during winter. This filter effect depends on the mass mean diameter (MMD) of the aerosol components. Particles with MMD of about 1 μm show a minimum in filter efficiency. The concentration and flux rates of trace substances in rainwater below the canopy are higher than above the canopy. Below the spruce canopy the throughfall concentrations were higher by a factor of 1.1 - 4.6 than below the beech canopy. The soil of the spruce forest received rainwater which had a free proton concentration which was by a factor of 4.8 ± 1 higher than the concentration above the canopy. In summertime the free acidity in rainwater reaching the beech forest soil was less than in the rainwater reaching the canopy. In assessing the effect of acid input on the soil of a spruce forest the throughfall has to be investigated and not the rainfall in the open field. The high free acidity of throughfall under the spruce canopy could be attributed to dry deposition especially gaseous dry deposition of SO_2 for which deposition velocities were deduced. It is suggested not to concentrate solely on acid rain but also on dry deposition when studying effects of atmospheric trace substances on ecosystems and to clearify whether airborne trace substances interfere with plants by direct interaction with the foliage or by indirect interaction via the forest soil.

6. ACKNOWLEDGEMENTS

Prof. B. Ulrich and Prof. H. Ellenberger (Göttingen) and Prof. R. Mayer (Kassel) placed the field site at our disposal and Mrs. G. Aheimer, Mr. K.P. Müller and Mr. H. Franken (Jülich) supported the work by continuous analytical and technical assistance. We are grateful for their kind help. The work was partly funded by the Federal Environmental Service (Umweltbundesamt), Berlin.

REFERENCES

Arwik, J.H.   1974   Barriers to the foliar uptake of lead
Zindahl, R.L.   J. Env. Qual. 3, 248-300

Davies, T.D.   1982   Dry deposition velocities of aerosol
Nicholson, K.W.   sulphat in rural eastern England,
                         H.W. Georgii and J. Pankrath (eds),
                         D. Reidel Publ. Comp. 31-42

Ellenberg, H.   1971   Integrated Experimental Ecology,
                         Springer Verlag, Berlin

Ernst, W.H.O.   1980   Personal communication

Georgii, H.W.   1980   Untersuchung über die trockene und
Gravenhorst, G.   feuchte Deposition von Luftverunrei-
Perseke, C.   nigungen in der Bundesrepublik
Rohback, E.   Deutschland, Bericht des Forschungs-
                         projektes 10402600 des Umweltbundes-
                         amtes, Berlin, Institut für Meteoro-
                         logie und Geophysik, Frankfurt/M.

Heinrichs, H.   1980   The role of forest vegetation in the
Mayer, R.   biogeochemical cycle of heavy metals
                         J. Env. Qual. 9, 111-118

Höfken, K.D.   1981   Untersuchungen über die Deposition
Georgii, H.W.   atmosphärischer Spurenstoffe an
Gravenhorst, G.   Buchen- und Fichtenwald, Berichte des
                         Instituts für Meteorologe und Geo-
                         physik der Universität Frankfurt/M.,
                         Nr. 46

Marggrander, E.   1982   Dry deposition of particles: com-
Flothmann, D.   parison of published experimentals
                         results with model predictions, in:
                         Deposition of atmospheric pollutants,
                         H.W. Georgii and J. Pankrath (eds),
                         D. Reidel Publ. Comp., 23-30

Marple, V.A.   1976   Impactor design
Willeke, K.   Atm. Env. 10, 891-896

Matzner, E.   1981   unveröffentlichte Daten

Mayer, R.   1982   Calculation of deposition rates from
Ulrich, B.   the flux balance and ecological
                         effects of atmospheric deposition upon
                         forest ecosystems, in: Deposition of
                         atmospheric pollutants, H.W. Georgii
                         and J. Pankrath (eds), D. Reidel
                         Publ. Comp., 195-200

Müller, K.P.   1982   The influence of immediate freezing on
Aheimer, G.   the chemical composition of rain-
Gravenhorst, G.   samples, in: Deposition of atmospheric

pollutants. H.W. Georgii and J. Pankrath (eds), D. Reidel Publ. Comp.,
125-132

Nürnberg, H.W.   1982   Wet deposition of toxic metals from
Valenta, P.   the atmosphere in the Federal Republic
Nguyen, V.D.   of Germany, in: Deposition of atmos-
                           pheric pollutants, H.W. Georgii and
                           J. Pankrath (eds), D. Reidel Publ.
                           Comp., 143-157

Sehmel, G.A.   1974   Predicted dry deposition velocities,
Hodgson, W.H.   Proc. of a Symp. held at Richland,
                          Wash. ERDA Symposium Series, CONF
                          740921 (1976), 399-422

Ulrich, B.   1979   Deposition von Luftverunreinigungen
Mayer, R.   und ihrer Auswirkungen in Waldöko-
Khana, P.K.   systemen in Solling, Schriften aus der
                          Forstlichen Fakultät der Universität
                          Göttingen und der Niedersächsischen
                          Forstlichen Versuchsanstalt, Bd. 58,
                          Frankfurt/M.

Ulrich, B.   1976   Input, output und interner Umsatz von
Mayer, R.   chemischen Elementen bei einem Buchen-
Khana, P.K.   und einem Fichtenbestand, Verhandlun-
Seekamp, G.   gen der Gesellschaft für Ökologie,
Fassbender, H.W.   Göttingen, 17-28

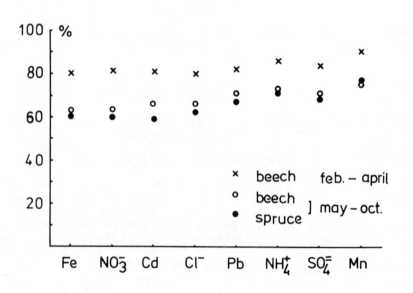

Fig. 1   aerosol concentrations below the canopy in a
          spruce and beech forest (Solling) expressed
          in % of the concentration above the forest.

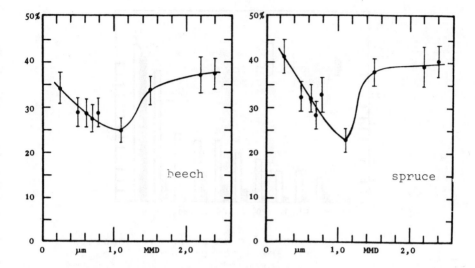

Fig. 2   reduction of aerosol concentrations by a forest
          canopy (expressed in % of the concentration above
          the canopy) as a function of the mass mean dia-
          meter of the component

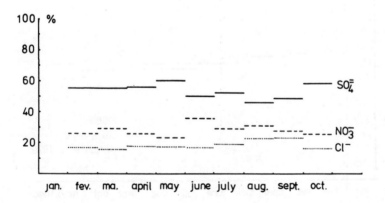

Fig. 3   The relative   contribution (in equiv-
          alent/l) of $SO_4^=$, NO_3^- and Cl^- to the sum
          of $SO_4^=+NO_3^-+Cl^-$ in rainwater above the
          forest (Solling, 1980)

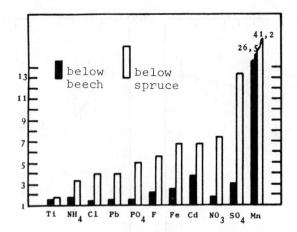

Fig. 4a concentration of rainwater constituents below a forest canopy (trough-
fall) expressed as a multiple of the
concentration above the canopy
(Solling, Feb.-Oct. 1980)

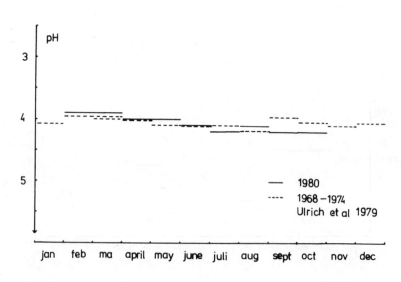

Fig. 4b ph values (volume averaged) for rainwater
sampled in the open field at the same site
during different periods

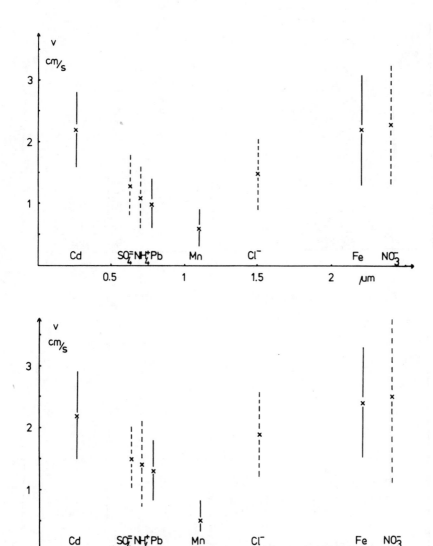

Fig. 5-6 dry deposition velocity of aerosol components
for a spruce forest (Fig.5 Feb.-April 1980,
Fig.6 May-October 1980) derived from precipitation chemistry above and below the canopy
and from impactor measurements. The dashed
vertical bars represent values interpolated
from aerosol size distribution measurements
and concentration differences of aerosols
above and below the canopy.

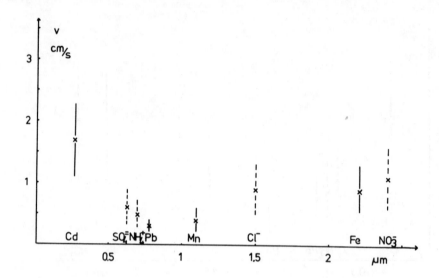

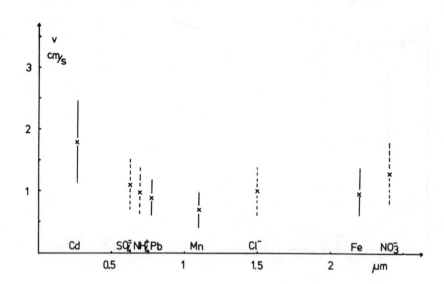

Fig. 7-8 dry deposition velocity of aerosol components
for a beech forest (Fig.7 Feb.-April 1980;
Fig.8 May-October 1980) derived from precipitation chemistry above and below the canopy
and from impactor measurements. The dashed
vertical bars represent values interpolated
from aerosol size distribution measurements
and concentration differences of aerosols
above and below the canopy.

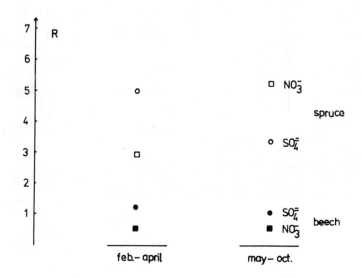

Fig. 9   The ratio R of gaseous dry deposition
and particulate dry deposition for
$SO_4^=$ and NO_3^- both of which can be
formed on the canopy by absorption of
gaseous precursors and be deposited
onto the canopy as particulate components.

	F^-	Cl^-	NO_3^-	$SO_4^=$	NH_4^+	Pb	Cd	Fe	Mn
beech									
feb.-april	–	20	18	16	14	18	19	19	10
may-oct.	29	34	37	29	27	29	34	37	25
spruce									
may-oct.	32	38	40	32	28	33	41	39	23

Tab. 1 concentration difference of aerosol components above and below a forest canopy (in % of the concentration above the canopy

	Cd	F^-	$SO_4^=$	NH_4^+	Pb	Mn	Cl^-	Fe	NO_3^-
μm	0.26	0.49	0.63	0.70	0.78	1.1	1.5	2.2	2.4

Tab. 2 mass mean diameter (MMD) of aerosol components measured may - october 1980 above the forest (Solling) by means of a 4 stage impactor

	Cd	Mn	Pb	Fe
beech				
total	3.8	26.5	2.6	2.5
dissolved	3.4	30.3	1.4	1.5
insoluble	16.5	2.0	2.5	3.0
spruce				
total	6.7	41.2	3.9	6.7
dissolved	4.4	46.7	2.5	3.1
insoluble	27.0	2.0	7.0	7.9

Tab. 3 concentration of trace-constituents in rainwater below the canopy (expressed as a multiple of the concentration above the canopy).

	Cl	NO_3^-	$SO_4^=$	NH_4^+	Fe	Mn	Cd	Pb	Ti
beech									
feb.-april	0.04	0.3	0.5	0.5	0.8	$<$4.0	2.1	0.8	-
may-oct.	0.2	0.4	1.3	0.3	1.0	$<$3.3	2.0	1.9	0.3
spruce									
feb.-april	0.6	3.3	4.0	0.9	2.0	$<$4.5	2.8	3.1	-
may-oct.	0.8	2.9	4.2	0.7	2.5	$<$9.3	2.5	2.7	0.2

Tab. 4 ratio dry deposition (gaseous plus particulate) wet deposition determined from concentration differences in rainwater above and below the canopy.

FIELD MEASUREMENTS OF NO and NO₂ FLUXES TO AND FROM THE GROUND

G. GRAVENHORST and A. BÖTTGER
Laboratoire de Glaciologie et Géophysique de l'Environnement
2 Rue Très Cloitres, F 38031 Grenoble

Summary

Acidity in soils may be destroyed or produced by fluxes of gaseous NO and NO₂ (NO_x) between the atmosphere and the soil. These NO_x-fluxes also play an important role in the atmospheric budget of NO_x and in assessing the natural and anthropogenic contribution to atmospheric NO_x-transferrates. We tried to determine NO_x flux rates at the earth-atmosphere interface. An exposure chamber was tilted over various surfaces in the field and used as a dynamical system to determine NO and NO₂ concentration changes in ambient air passing through the chamber. These concentration changes, measured by a chemi-luminiscent detector, were used to derive deposition velocities or source strengths. The determined NO₂-deposition velocities for water and snow surfaces (<0.015-0.02 cm s⁻¹) and forest floors (about 0.015-0.04 cm s⁻¹) are smaller than for pastures, meadows, lawn and bare soil by up to a factor of ten. The NO₂ deposition velocity for a cement surface under ambient conditions ranged between 0.05 and 0.09 cm s⁻¹. The NO-deposition velocities for water, concreate and lawn were less by a factor of about 10 than the NO₂-deposition velocities. From lawn, pastures, meadows and forest floors NO was usually volatilized at a rate of 1-10 µg N m⁻²h⁻¹. At a special site where the NO-concentration fluctuated considerably between 2 and 160 ppb bare garden soil volatilized as well as absorbed NO. Comparing NO-volatilization from and NO₂-deposition to the same soils, here investigated, the net effect is generally a NO_x-flux towards the ground at ambient NO₂-concentrations larger than 1 ppb.

1. INTRODUCTION

In order to describe the atmospheric fate of nitrogen oxides and acidic inputs to various ecosystems the NO and NO₂ fluxes across the earth-atmosphere inter-face have to be known. In budget considerations (Robinson and Robbins, 1968, 1970a, 1970b, 1971, 1975; Söderlund and Svensson, 1976; Söderlund, 1976; Bonis et al., 1980; Galbally, 1980) the NO₂-transferrates to the ground were estimated by assuming a representative NO₂ concentration in the atmospheric boundary layer close to the ground and NO₂-deposition velocity between 0.2 and 1.0 cm s⁻¹. These values were often chosen in reference to SO₂.

Measured NO₂ uptake rates are, however, very seldom reported. In an exposure chamber the NO₂ concentration in air passing through the chamber decreased when plants were present (Hill, 1971; Bennett and Hill, 1975). The suggested NO₂ uptake by plants was more pronounced during day time than in the dark. We deduced an average daily NO₂-deposition velocity per floor area of about 0.4-0.6 cm s⁻¹ from their chamber measurements. A similar influence of light conditions on the inter-action of NO₂ with selected plant species are reported by Rogers et al. (1977) and Rogers et al. (1979) and with spruce needles by Greenfelt (1982). The determined deposition velocities per needle area were 0.065-0.17 cm s⁻¹ during days and of

0.028-0.08 cm s^{-1} during nights. The reported data for soybean and corn were converted by us to deposition velocities per floor area. The thus estimated NO_2-deposition velocities range from about 0.1 cm s^{-1} for soybean to about 0.02-0.04 cm s^{-1} for cornplants without significant dependance on NO_2 concentration. The deduced deposition velocity for 12 day old cornplants increased from about 0.02 cm s^{-1} to 0.2 cm s^{-1} when the light level was increased from 0 to 343 μ einstein m^{-2}s^{-1}. The increased NO_2 uptake rate on illumination in both investigations was attributed to additional NO_2-absorption during photosynthesis by inner leaf surfaces mediated primarily by physical exchange processes. Although the residence time of air molecules in the exposure chamber was of the order of 20 min photolysis effects in air were suggested to be very small compared with NO_2 uptake by plants (Rogers et al., 1979).

A diurnal variation could, however, not be detected over a tree stand by NO_2 gradient measurements, which were interpreted to be caused by a NO_2 uptake of the canopy with a NO_2 deposition velocity of 1.4 $\pm$ 9.5 cm s^{-1} in summer and 5.9 $\pm$ 4.5 cm s^{-1} in winter (Kessler, 1979). Over the same tree covered ground no systematic NO_2-gradient in a 100 m thick layer above the trees could be detected in another time period (Platt, 1977). NO_2 gradients over a vertical distance of about 100 m can be caused by an inhomogeneous distribution of gaseous constituents influencing the NO_2 concentration. The residence time of inert air molecules in the boundary layer between 20 and 120 m was at that site about 3 min determinded from average turbulent transfer velocities of about 0.5 m s^{-1} during 256 measurements throughout the year (Platt, 1977). This residence time is long enough compared to characteristic times for NO_2-production and destruction mechanisms in the gasphase so that NO_2-concentration gradients can be produced which are not caused by NO_2 absorption at the ground. Furthermore the surface structure and NO_x-distribution neighboring the measuring site could be quite inhomogenous with respect to the covered hight intervall during the gradient measurements that NO_2-gradients in the boundary layer may result without the influence of NO_2-absorption at the ground. Between 0.2 and 2 m above the ground no significant NO and NO_2 gradient could be detected by Garland and Cox (1977), indicating NO_2-deposition velocities smaller than ca. 0.4 cm s^{-1} at this site. In a field study, similar NO_2 deposition velocities were found (Varhelyi, 1980).

In a laboratory study the influence of a bare soil on the NO_2-gasphase concentration has been compared with the effect on ethylene and sulfur dioxide (Abeles et al., 1971). The NO_2 uptake rate seemed to be slower by a factor of about 60-100 than the SO_2 uptake rate. This would suggest a NO_2 deposition velocity of about 0.01 cm s^{-1} for these conditions. Higher values (0.3-0.6 cm s^{-1}) for NO_2 absorption by soils were reported by Böttger et al. (1978). They took the soil into the laboratory thus disturbing the soil structure and greatly enlarging the surface area of the soil exposed to the atmosphere. Similar high NO_2 deposition velocities were determined for freshly produced cement and soil surfaces (Judeikes and Wren, 1978).

The absorption of atmospheric NO_2 by a flat rainwater surface was first determined by Georgii (1963). From the reported and additional information we could deduce a NO_2-deposition velocity of about 0.01 cm s^{-1} for rainwater and destilled water which had been in contact with the atmosphere for several days during the measurements. Such a slow NO_2 uptake by water was also found by Beilke (1970) studying the interaction of SO_2 and NO_2 with artificial rain. The first quantitave values for NO_2 deposition velocities for water surfaces and atmospheric NO_2 concentration were given by Böttger et al. (1978) as 0.015 cm s^{-1} for seawater and 0.01 cm s^{-1} for tab water. The small difference in uptake rates found for rainwater and destilled water (Georgii, 1963) and for seawater and tab water (Böttger et al., 1978) suggests that the chemical composition of atmospheric water does not greatly effect the uptake rate of NO_2.

The absorption of gaseous NO_2 by liquid water should, however, not proceed as a first order reaction with respect to NO_2 (Lee and Schwartz, 1981). To quantify a NO_2 flux to wet surfaces - and surfaces exposed to the atmosphere are generally covered with a liquid film - by means of a NO_2-deposition velocity, the value of which is taken to be indepedant of the NO_2 gasphase concentration, is therefore questionable, especially when measurements are performed at NO_2-concentrations which are high compared to atmospheric conditions (e.g. Ghiorse and Alexander, 1976; Prather et al., 1976a). The resulting effects may be misleading. Within the range of encountered atmospheric NO_2 concentration at one special site the conceptual use of a NO_2-deposition velocity to quantify NO_2 transferrates from the atmosphere to the ground seems, however, a justified first step in estimating NO_2 flux rates. We tried, therefore, to determine NO_2-deposition velocities for various surfaces in the field. Because of the limited nature of this explorative study it was not aimed at determining uptake mechanisms, reaction pathways, and special surface parameters but rather to get an impression of the order of magnitude of NO_2-deposition velocities to various surfaces.

Some of the surfaces investigated, did, however, not only absorb NO_2 but also volatilized NO. Gaseous NO and NO_2 may escape from the soil into the atmosphere following the chemical transformation of nitrite according to

$$2 H^+ + 3 NO_2^- \rightarrow 2 NO + NO_3^- + H_2O \quad (1)$$

$$H^+ + 2 NO_2^- \rightarrow NO + NO_2 + OH^- \quad (2)$$

This chemical formation of gaseous NO and NO_2 from nitrite represents only a small fraction of the gaseous N-loss from soils by biological denitrification of nitrate and nitrite and nitrification of ammonium to either molecular nitrogen and/or nitrogen oxides (NO, NO_2, N_2O) (Allison, 1966). Nitrite can be reduced by several metallic ions. Kinetic data are given for Mo(V), Pu(III), Fe(II) and As(III) by Bamford and Tipper (1972) The concentrations seem, however, to be too low in soils to be effective (Nelson and Bremner, 1970). Although the last step in this NO formation is a non biological process (Tyler and Broadbent, 1961) the reservoir of NO_2^- as a precursor of NO will expire unless the consumed NO_2^- is replaced by biological processes. The release rate of NO and NO_2 from leaves of herbizide treated soybeans, for example, was governed by the disturbance of the metabolic balance of the plants (Klepper, 1979). For thermodynamical reasons the pathway (1) for the nitrite reduction should dominate (Van Cleemput and Baert, 1976), whereas some experiments are believed to be more in accordance with the pathway (2) (Nelson and Bremner, 1969). In some laboratory measurements (Smith and Clark, 1960; Tylor and Broadbat, 1960) no NO production could be found. It was explained by Nelson and Bremner (1969) that in a closed system liberated NO may again be absorbed by the soil. The main soil parameters increasing a NO_x-volatilization seem to be a low pH value at the active sites (Gerritsen and de Hoop, 1957; Bremner and Nelson, 1968; Nelson and Bremner, 1969; Van Cleemput et al., 1975) and high organic matter content (Bremner, 1976).

From an unfertilized soil (pH 5.0-5.2) 0.4 kg N $m^{-2} a^{-1}$ was volatilized as NO_2 (Makarov, 1969). An average NO source strength twice as strong was found in Australia (Galbally and Roy, 1978). The estimated NO_x volatilization from mineral nitrogen fertilizer application was estimated to contribute only several percent to the total NO_x-flux rate into the atmosphere (Böttger et al., 1978). NO volatilization and NO_2 absorption can eventually counteract each other with respect to changes in the nitrogen pool of the atmosphere as well as in the soil. Since for both fluxes only very rare field data exist we tried at random to determine flux rates between the atmosphere and the ground at some sites in Germany.

2. EXPERIMENTAL SETUP

NO and NO_2 transferrates between the ground and the atmosphere were deduced from NO- and NO_2 concentration changes in air being in contact with the ground. The NO and NO_2 concentration of outside air was measured by NO chemiluminescent detectors before and after having passed through a plexiglass chamber which was tilted over the soils in the field. Such exposure chamber measurements have in general been applied to investigate gas-soil and gas-plant interactions in a static or dynamical mode (Hill, 1967; McGarity and Rajaratnam, 1973; Payrissat and Beilke, 1975). The chamber used here had a volume of 9 l with a bottom area of 0.088 m². It was lozenge-shaped to facilitate a continous flow within the chamber. The chamber could be moved horizontally from the side over the ground to be investigated and lowered onto a metal frame by a motordriven lever-arm in a variable time intervall. The metal frame with a sharp blade at its bottom was pressed about 5 cm into the soil once at each measuring site to prevent uncontrolled air exchange with the surroundings. The upper part of the frame was U-shaped and filled with water to ensure no leakages. When the interaction of a cement surface was investigated the plexiglass chamber itself was set on the floor and sealed with a mouldable plastic. The plexiglass chamber was lowered onto a water filled tank of the same size and shape to follow up NO and NO_2 concentration changes in air streaming over the water surface. The filtered outside air was usually sucked through the system at a flow rate of ca. 500 l h^{-1}. Large differences in NO production rates which were only due to changes in flow rates could not be detected (Fig. 1). Since sometimes differences between the incoming and the outgoing flow rates were observed, the air was also pushed through the system significantly different results could, however, not be detected for these two operation modes.

The average residence time of air molecules in the exposure chamber was about 70 s. By flushing the chamber continuously when lowered to the soils the changes of factors which affect plants and soils like CO_2 concentration, temperature and water vapour concentration were kept small compared to a closed system of a batch reactor. In order to reduce the disturbance of the normal soil conditions still further the chamber was intermittantly lifted to enable adjustment to the environment and lowered to measure fluxes for half an hour. When lowered the NO and NO_2 concentrations were monitored for five minutes alternating at the inlet and outlet of the chamber. The differences between these two air sampling points were attributed to the interaction of the air with the soil taking into account blank runs in which the exposed soil was covered by a plastic floor. A deposition velocity v was deduced according to:

$$v = \frac{F}{A} \times \frac{\Delta c}{c} \qquad (3)$$

(F: air flow rate through the chamber, A: area of the chamber floor, Δc: concentration difference between ingoing c_1 and outgoing c_2 concentration, c: 0.5 ($c_1 + c_2$).

The exposure chamber is therefore not interpreted as a continous stirred tank reactor with the same concentration anywhere in the chamber (Jeffries et al., 1976). A production rate P was correspondingly calculated according to

$$P = \frac{F}{A} \times \Delta c \qquad (4).$$

3. DISCUSSION

The use of such a chamber to determine NO and NO_2 fluxes between the soil and the atmosphere introduces changes within the system under observation. This influence may strongly affect the measured gasphase concentration changes to be only attributed to the interaction between NO_x and the soil system. For example, the light intensity in the photoactive wavelengths will be reduced by the chamber and the ozone molecules will be destroyed at the surfaces within the chamber. Various homogeneous gasphase reaction rates are thus altered. The newly introduced production and destruction rates may change the NO_x-concentrations in the air and thus mask the result of NO and NO_2 interaction between the soil and the air. The characteristic time to reach steady state conditions for NO, NO_2 and O_3 in an undisturbed air volume is of the order of two minutes when only these three reactions are considered:

$$NO_2 + h\nu \rightarrow NO + O \qquad (5)$$
$$O + O_2 + M \rightarrow O_3 + M \qquad (6)$$
$$NO + O_3 \rightarrow NO_2 + O \qquad (7)$$

The residence time of air molecules in our chamber was of similar duration. It should, therefore, be difficult to separate the direct influence of the surfaces on the NO and NO_2 concentrations from the indirect influences of the measuring system.

Three experimental results indicate, however, that the measured NO and NO_2 concentration differences can be mainly attributed to the direct influence of the soil: 1. the observed NO-volatilization rate did not differ significantly when outside air and synthetic air streamed through the chamber, 2. the NO_2 deposition velocities did not differ systematically for day and night time conditions, 3. the NO concentration in air passing through the chamber did not change although O_3 and NO_2 concentration decreased due to dry deposition in blank runs and on cement grass, and water surfaces.

At the measuring site in Köln-Fittard the O_3 concentration was monitored from 5.10.82 to 28.10.82. The ambient ozone concentration in some centimeters above the ground was most often only several ppb. The reaction on the ground and with atmospheric NO can be responsible for this low O_3 level. These O_3 concentrations are in accord with the results of a station monitoring ambient air quality near Godorf, about 20 km south of our measuring site (Deimel, 1982). The O_3 concentrations in about 10 m above ground averaged 7.6 $\pm$ 7 ppb during daytime (0700-1800) and 5.8 $\pm$ 7 ppb during night time (1800-0700) conditions in our measuring periods. The interference of the exposure chamber with O_3 should, therefore, have a minor effect on the NO_x concentration changes within the chamber. During the few occasions when higher O_3 concentrations were measured the O_3 concentration decrease between inlet and outlet indicated an O_3 deposition velocity for water of about 0.1 cm s^{-1} and for grass of about 0.2 cm s^{-1}. The O_3 deposition velocities fall in the upper range reported e.g. by Galbally and Roy (1980).

3.1 NO_2-FLUXES

The NO_x deposition velocities and source strengths derived for various surfaces are summarized in Tab. 1. The deposition velocities derived for NO_2 seem to be noticable lower than reported in other investigations for SO_2. This indicates that an analog treatment of both gases with respect to dry deposition is not justified. Generally the NO_2 deposition velocities fall below 0.1 cm s^{-1} for rather smooth surfaces (bare soil, concrete, water) and between 0.1 and 0.2 cm s^{-1} for rougher surfaces (lawn, meadow, pastures). The lowest values were found for a snow surface ($v < 0.015$ cm s^{-1}), the highest value for a meadow with a relatively high canopy ($v = 0.4$ cm s^{-1}). The smaller NO_2-deposition velocities may be caused by both a smaller effective surface area as well as a less effective surface reactivity. The deposition velocity for water surfaces falls in the same range as found by Böttger et al. (1978). If the enhanced NO_2 absorption rates for plants under illumination (Hill, 1971; Rogers et al. 1979; Greenfelt, 1982) are due to plant uptake through stomata the NO_2 absorption by inner leave surfaces, the plant cell, or the cell solute with their metabolisms must have properties greatly increasing the NO_2-uptake compared to other system such as rainwater, seawater or wetted soils. In our limited field experiments we could, however, not detect a significantly enhanced NO_2 deposition velocity during daytime for grass surfaces. To substantiate this result further field studies with green plants will be conducted in the next growing season.

The high deposition velocities (0.3-0.8 cm s^{-1}) reported for freshly prepared surfaces of sandy loam soil, adobe clay soil and cement (Judeikes and Wren, 1977) were not encountered for a cement surface exposed in front of a garage to ambient atmospheric conditions in a residential area close to the city of Köln. This is in agreement with the finding that the initially measured deposition velocities on freshly produced soil and cement surfaces decreased gradually with time and approached the detection limit of the experimental setup. In light of the values for bare soils in the field (0.05-0.1 cm s^{-1}) the laboratory measurements of $v(0.3$-0.6cms^{-1}) by Böttger et al. (1978) indicate that the disturbance of the soils affected their absorption capacity probably by increasing the effective surface area.

The large variability of the NO_2 deposition velocity at one site can be seen in Fig. 2. The mean half hour values ranged from < 0.015 to 0.19 cm s^{-1} during a time period which was characterized by frequent snow and rainfall. Snow reduced the NO_2 deposition velocity to less than 0.015 cm s^{-1}. During snow melting v increased again. A clear effect of rain on the deposition velocity can, however, not be distinguished from other influences. Rain can either restore an effective absorption surface by washing off absorption prohibitors and activating new sites or destroy active sites by covering the surface with a liquid film.

The variation of v in Fig. 2 is probably due to changes in the properties of the absorbing surface. In reality additional fluctuations in v may be introduced by changes in the turbulent and molecular diffusion resistance, although the small deposition velocities suggest that the turbulent NO_2 transport to the ground is not the rate limiting step in NO_2 absorption.

The general application of deposition velocities determined in such a flow system to larger areas is quite prolematic. The earth surface is usually not as smooth and homogeneous as investigated here and the quantitative influence of spatial and temporal variations in effective absorption area and roughness height, soil conditions and atmospheric composition is not known. A different approach than here persued to estimate the gaseous dry deposition to the real earth surface integrating over the various small and large scale processes is urgently needed.

3.2 NO FLUXES

The NO concentration in air passing through the exposure chamber usually increased (Tab. 1). Only over water, cement and sometimes grass no NO volatilization could be measured. The found NO deposition velocities to water and cement are generally smaller by a factor of ten than the NO_2-deposition velocities. From lawn, meadows, pastures and forest floors about 1-10 μg N m^{-2}h^{-1} are volatilized as NO. This range is about half the average flux rate reported by Galbally and Roy (1978). More NO seems to be released from the forest floor which was fertilized in July 1973 with primarily a high amount of nitrogen (266 kg/ha) and in October 1975 mainly with calcium (1188 kg/ha). The two forest sites are extensively described Ellenberg (1971). Because of the low pH value of the investigated beech and spruce floor (pH 3.5-4.5) the conditions for NO volatilization seem to be favourable. N_o

systematic diurly variation of NO volatilization from the forest floors could be detected. The canopy protects the underlying ground from large environmental variations so that longer measuring periods will be needed to detect such changes. An increased NO volatilization when sun radiation reached the ground indicates that increase in temperature can enhance the NO release. Direct nitrite photolysis as found by Zafiriou and True (1979) in seawater does not seem to be a main NO production mechnism since it should result in a pronounced diurnal variation.

The soil, however, does not seem to always liberate NO. At a site in a residential area in the vicinity of large NO_x sources (traffic and industry) a garden soil did not only volatilize but also absorbed NO. The past treatment of this garden soil is not known. It may have been fertilized some time ago by the owners. In Fig. 3 the measured fluxes between atmosphere and this soil are plotted as a function of ambient NO concentration. It seems that at lower NO concentrations NO is emitted whereas at higher ambient NO concentration the soil seems to have the ability to absorb NO. A certain distinct threshold NO concentration separating source and sink functions is not apparent although NO concentrations around 60 ppb may indicate such a transition. Other influences like hystereses effects, atmospheric pressure changes, analytical interferences with unknown compounds emanating from the soil pores or present in the ambient air can prohibit to isolate the effect of one parameter. At all other sites in Germany, no NO absorption was detected probably because of the low NO concentrations encountered at these rural sites. Some threshold value is, however, also suggested by measurements using our dynamical system as a static chamber. When NO was released from the soil the NO concentration in the closed chamber increased only to a certain saturation level indicating that either the NO release from the soil pores went to zero or NO was increasingly destroyed finally balancing its production. A similar equilibrium NO concentration seemed to be reached in measurements by Galbally and Roy (1978). An NO uptake by the soil could also be detected by them by introducing higher initial NO-concentrations than certain equilibrium values. The downward NO fluxes allow the calculations of a NO deposition velocity to the ground. It turned out to be of the order of 0.02 ± 0.015 cm s^{-1}. The conditions under which these values characterize a NO transfer from the atmosphere to the soil and when the NO sink turns into a NO source are, however, not known. The behaviour of NO at the atmospheric interface is therefore another good example that processes relevant for atmospheric conditions do not allways proceed by first order kinetics. Results determined at NO levels high compared to atmospheric concentrations (Prather et al., 1976b) are, therefore, primarily restricted to the conditions of the measurements.

It seems that generally NO is released from the soil and NO_2 absorbed. The net effect of these two fluxes on the N budget of soils depends on a variety of parameters. At high NO_2 concentration levels in ambient air ($NO_2 > 1$ ppb) the net flux between soil and atmosphere will probably be directed to the ground. In remote areas a net NO_x emission into the atmosphere can result from soils when high concentrations of organic matter in soil, low soil pH values at active sites and high temperature favour NO volatilization. When high amounts of N-fertilizer are applied a NO volatilization from the soil may counter balance the NO_2 deposition to the soil.

Analytical instruments were kindly put to our disposal by W. Fricke, Institut für Meteorologie and Geophysiks, Frankfurt, U. Lenhard, Institut für Physikalische Chemie, Kiel and Dr. M. Deimel, Amt für Lufthygiene der Stadt Köln. Meteorological data were supplied by Ch. Koch, German Weather Service, Essen. We gratefully acknowledge their help.

REFERENCES

Allison, F.E.	1968	The fate of nitrogen applied to soils in : Advances in Agronomy, 18, A.G. Norman (Ed.), Academic Press, New York and London, 219-258
Bamford, C.H. Tipper, C.F.H.	1972	Chemical Kinetics, Vol. 7, Elsevier Publ. Comp. Amsterdam, London, New York, 471-473
Bennett, J.H. Hill, A.G.	1975	Interactions of air pollutants with canopies of vegetation, in: Responses of plants to air pollution D.M. Mudd, T.T. Kozolowski (eds), Academie Press, Chap. 12
Böttger, A. Ehhalt, D.H. Gravenhorst, G.	1978	Atmosphärische Kreisläufe von Stickoxiden und Ammoniak, Berichte der Kernforschungs- anlage Jülich, JÜL 1558, ISSN 0366-0885
Bonis, K. Meszaros, E. Putsay, M.	1980	On the atmospheric budget of nitrogen compounds over Europe, Jdöj aras 84, 2, 57-68
Bremner, J.M.	1976	Role of organic matter in volatilization of sulfur and nitrogen from soils Intern. Symp. on Soil organic matter studies Braunschweig, F.R.G., Paper No. IAEA-SM-211/9
Bremner, J.M. Nelson, D.W.	1968	Chemical decomposition of nitrite in soils Trans. 9th Int. Cong. of soil science, Adelaide, Aug. 1968, 2, 1968, 495-503
Deimel, M.	1982	Tagesprotokolle der Mess-Station Godorf, Amt für Lufthygiene der Stadt Köln, unver- öffentlichte Daten
Ellenberg, H.	1971	Integrated Experimental Ecology Springer Verlag, Berlin
Galbally, I.E. Roy, C.R.	1980	Destruction of ozone at the earth surface, Quat. J. Roy. Met. Soc. 106, 599-620
Galbally, I.E. Freney, J.R. Denmead, O.T. Roy, C.R.	1980	Processes controlling the nitrogen cycle in the atmosphere over Australia, in: Biogeochemistry of acient and modern Environments, P.A. Trudinger and M.A. Walter (eds), Australian Academy of Science and Springer Verlag
Galbally, I.E. Roy, C.R.	1978	Loss of fixed nitrogen from soils by nitric oxide exhalation Nature 275, 734-735

Garland, J.A. Cox, L.C.	1977	Dry deposition of nitrogen oxides Environmental and Medical Sciences Division, Progress report, January-December 1976, AERE Harwell, Oxford, p. 89
Georgii, H.W.	1963	Oxides of Nitrogen and Ammonia in the Atmosphere, J. Geophys. Res., 68, 13, 3963-3970
Ghiorse, W. Alexander, M.	1976	Effect of microorganisms on the sorption and fate of sulfur dioxide and nitrogen dioxide in soil, J. Environ. Anal. 5, 3, 227-230
Greenfelt, P.	1982	Dry deposition of nitrogen oxides, paper presented at EC-COST 61a bis, 3. Meeting Working Parties 4 and 5, Berlin, Sept. 7-9, 1982
Hill, A.C.	1967	A special purpose of plant environmental chamber for air pollution studies, J. Air. Poll. Control Ass., 17, 11, 743-748
Judeikis, H.S. Wren, A.G.	1978	Laboratory measurements of NO and NO_2 depositions onto soil and cement surfaces Atmospheric Environment, 12, 2315-2319
Kessler, C.	1979	UV-Spektroskopische Bestimmung der trockenen Deposition von SO_2 und NO_2 mittels der Gradientenmethode, Diplomarbeit, carried out at the Kernforschungsanlage Jülich, Institut für Atmosphärische Chemie
Klepper, L.	1979	Nitric oxide (NO) and nitrogen dioxide (NO_2) emissions from herbicide-treated soybean plants Atm. Env. 13, 537-542
Lee, Y.N. Schwartz, S.E.	1981	Reaction kinetics of nitrogen dioxide with liquid water at low partial pressures J. Phys. Chem., 85, 840-848
Makarov, B.N.	1969	Liberation of nitrogen dioxide from soil, Soviet soil science, 20-25 translated from Pochvovedeniye No. 1, 49-53
Mc Garity, J.W. Rajaratnam, J.A.	1973	Apparatus for the measurement of losses of nitrogen as gas from the field and simulated field environments, Soil Biol. Biochem. 5, 121-131
Nelson, D.W. Bremner, J.M.	1969	Factors affecting chemical transformations of nitrite in soils, Soil Biol. Biochem. 1, 229-239

Nelson, D.W.   1970   Role of soil minerals and metallic cations in
Bremner, J.M.   nitrite decomposition and chemodenitrification
                         in soils, Soil Biol. Biochem, 2, 1-8

Platt, U.F.   1977   Mikrometeorologische Bestimmung der SO_2-Ab-
                       scheidung am Boden, Diss. Universität Heidel-
                       berg

Prather, R.J.   1976a   Sorption of nitrogen dioxide by calcareous
Miyamoto, S.   soils, Soil Sci. Soc. Amer. Proc. 37, 860-863
Bohn, H.L.

Prather, R.J.   1976b   Nitric oxide sorption by calcareous soils: I.
Miyamoto, S.   capacity, rate and sorption products in air
Bohn, H.L.   dry soils, Soil Sci. Soc. Amer. Proc. 37,
                          877-879

Robinson, E.   1968   Sources, abundance and fate of gaseous atmo-
Robbins, R.C.   spheric pollutants, Stanford Res. Institute,
                        Menlo Park, Cal. Final Report

Robinson, E.   1970a   Gaseous atmospheric pollutants from urban and
Robbins, R.C.   natural sources
                        In: Singer, S.F. (ed) Global effects of
                        environmental pollution, D. Reidel Publishing
                        Company, Dordrecht-Holland, 50-64

Robinson, E.   1970b   Gaseous nitrogen compound pollutants from
Robbins, R.C.   urban and natural sources
                        J. Air Pollut. Control Ass. 20: 303-306

Robinson, E.   1971   Sources, abundances, and fate of gaseous atmo-
Robbins, R.C.   spheric pollutants supplement, Supplement
                        Report prepared for American Petroleum
                        Institute, Stanford Research Institute, SRI-Pro-
                        ject PR-6755, API-Publ. No. 4015

Robinson, E.   1975   Gaseous atmospheric pollutants from urban
Robbins, R.C.   and natural sources, In: Singer, S.F. (ed) The
                        Changing Global Environment, D. Reidel
                        Publishing Company, Dordrecht-Holland,
                        111-123

Rogers, H.H.   1977   Measuring air pollutant uptake by plants:
Jeffries, H.E.   a direct kinetic technique, J. Air Poll. Contr.
Stahel, E.P.   Ass., 27, 12, 1192-1197
Heck, W.W.
Ripperton, L.A.
Witherspoon, A.M.

Rogers, H.H.   1979   Measuring air pollutant uptake by plants:
Jeffries, H.E.   nitrogen dioxide, J. Environm. Qual., 8, 4,
Witherspoon, A.M.   551-557

Tyler, K.B. Broadbent, F.E.	1960	Nitrite transformations in California soils, Soil Sci. Soc. Proc. 24, 279-283
Van Cleemput, O. Baert, L.	1976	Theoretical consideration on nitrite self - decomposition reactions in soils, Soil Sci. Soc. Am. J., 40, 322-324
Varhelyi, G.	1980	Dry deposition of atmospheric sulfur and nitrogen oxides, Idöjaras, 84, 2, 15-20
Zafiriou, O.C. True, M.B.	1979	Nitrite photolysis as a source of free radicals in productive surface waters Geophys. Res. Lett., 6, 2, 81-84

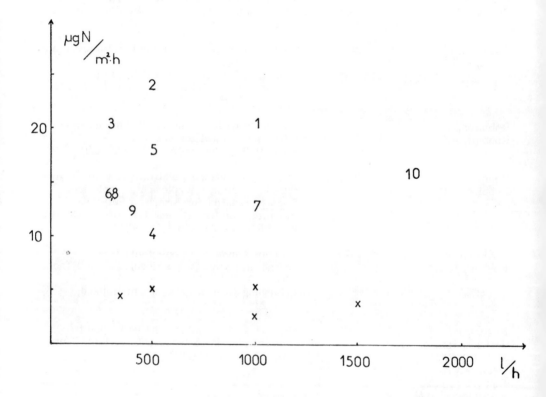

Fig. 1: NO source strengths (forest floor) determined at different flow rates. The numbers indicate the sequence of the measurements.

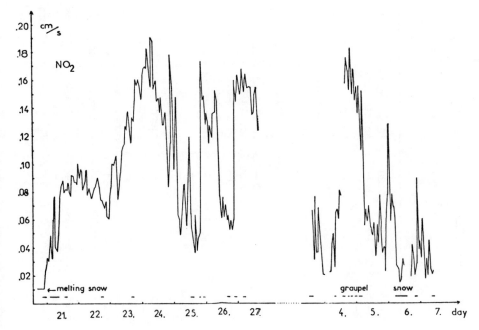

Fig. 2: Variation of NO$_2$ deposition velocities determined at one
site (lawn). The dashed horizontal lines indicate time of
precipitation.

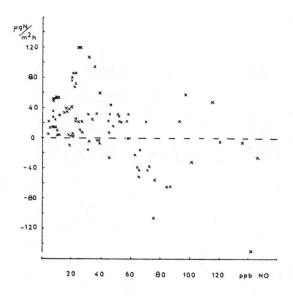

Fig. 3: The NO flux from and to a garden soil as a function of NO
concentration in ambient air.

TABLE I - NO and NO$_2$ flux measurements in the field

site	date	time	surface	NO ppb	NO$_2$ ppb	V_{NO_2} cm/s	NO source $\mu gNm^{-2}h^{-1}$	remarks
Kl. Feldberg	06.08.78	14.00-20.00	bare soil	2-4	10-15	0.1	20	blew barries
"	06.08.78	14.00-20.00	meadow	2-4	10-15	0.1 -0.3	0.9-1.2	acidic side
Volgelsberg	30.07.78	15.00-22.00	pasture	1-2	2-5	0.03-0.015	1-10	-
Solling	11.07.79	17.00-20.00	forest floor	2-5	-	-	0.9-2.0	beech
"	12.07.79	09.00-12.00	"	2-5	-	-	1.2-2.4	"
"	12.07.79	12.00-15.00	"	2-5	-	-	6-11	direct sun
"	13.07.79	18.00-23.00	"	2-5	-	-	0.9-1.6	beech
"	14.07.79	07.00-12.00	"	2-5	-	-	0.9-2	"
"	12.07.79	16.00-21.30	"	1-6	-	-	2-10	spruce
"	13.07.79	10.00-17.00	"	1-5	-	-	3-5	"
"	14.07.79	14.00-22.00	"	1-5	-	-	10-165 median 20	fertilized 1972 and 1975
Hoffnungsthal	20.07.79	15.00-20.00	pasture	4-10	-	-	1-2	pH 5.5-6.0
Flittard	21.07.79	10.30-14.30	stubble field	5-10	-	-	4-7	pH 6.0-6.5
"	25.07.79	08.00-18.30	cement	35	75	0.03-0.1	-	in front of garage
Solling	27.09.79	08.00-14.00	meadow	2-5	6-10	0.1-0.4	-	-
"	30.09.79	09.00-12.00	"	2-5	5-10	0.1-0.2	-	-
"	29.09.79	09.00-11.00	forest floor	2-6	5-15	0.01	-	spruce
"	29.09.79	12.00-14.00	"	2-6	5-15	0.03	-	beech
Flittard	01-05.11.80	continuously	lawn	-	15-30	0.09-0.25	-	garden
"	20-27.12.80	"	"	-	10-90	0.02-0.19	-	"
"	03-07.01.81	"	"	-	10-50	0.015-0.13	-	snow, rain
"	05-09.10.82	"	"	-	5-25	0.1-0.15	<1	"
"	24.10.82	11.00-17.00	cement	-	5-15	0.05-0.09	-	in front of garage
"	24-26.10.82	17.00-14.00	bare soil	2-86	5-25	<0.015-0.14	40±34	V_{NO} <0.01 cms^{-1}
"	26.10.82	15.00-18.00	tab water	15-30	7-12	0.015-0.02	-	V_{NO} 0.024 ± 0.014 cms^{-1}
"	26-28.10.82	18.00-02.00	bare soil	26-160	7-21	<0.017-0.09	27±12	V_{NO} <0.01 cm s^{-1} V_{NO} 0.02 ± 0.017 cms^{-1}

STUDIES OF THE ACIDITY AND CHEMICAL COMPOSITION OF CLOUDS

A.R.W. MARSH
Central Electricity Research Laboratories
Leatherhead, Surrey, United Kingdon

Summary

    Measurements of the composition of clouds have been made as part of a
programme of tracking and sampling power plant plumes over the North Sea.
CERL developed a centrifugal cloud water collector for these experiments.

    Most of the measurements were made in stratiform cloud with cloud
water loadings > 0.5 g/M^3. Concentrations in cloud are higher than in
precipitation for all ions and this is reasonable in terms of the growth
mechanism of cloud drops to form precipitation.
    For a range of SO_2 concentrations cloud water was found to contain
more sulphate than can be accounted for in terms of local SO_2 saturation.
    The composition of the cloud water samples from one flight have been
examined in detail to determine the sources of the ionic species. The
unusual meteorological conditions on this occasion apparently prevented any
major chemical transformation within the cloud drops. It is concluded that
57% of the cloud water acidity within the plume derived from HCl gas and
that the sulphate derived from acid ammonium sulphate aerosol, the
stoichiometry of which can be represented as 17 to 36% H_2SO_4 and 64 to 83%
$(NH_4)_2SO_4$.

1. INTRODUCTION

    Physical and chemical processes during the nucleation of natural
clouds and subsequently within cloud droplets, form an essential part of
the pathway linking atmospheric emissions to the composition of rain and
snow. At the simplest level the existence of clouds is an essential
requirement for rain or snow to occur. In addition, however, it has been
postulated that significant chemical transformation can take place in the
aqueous phase in clouds, particularly the oxidation of dissolved SO_2, but
this supposition is based on laboratory measurements of reaction rates and,
apart from one set of measurements which are rather inconclusive (Hegg and
Hobbs, 1981), no actual direct evidence for in-cloud oxidation has been
obtained.
    As part of the CERL programme of aircraft flights to track and sample
power plant plumes, cloud water was sampled and analysed on several
occassions. A description of the cloud water separator is given by
Walters, Moore and Webb, 1982. On one particular flight, 28 January 1981,
an area of stratiform cloud was sampled which coincided with the plumes
from the Yorkshire and Midlands Power Station complexes. The plumes were
sampled just off the East coast of the UK and corresponded to about 4 hours
of plume travel. A very important point in connection with these
measurements is that the rates of oxidation within the plume were very low
as evidenced by the high proportion of NO in the NO_x of the plume,
(~80%). A complete pH record was obtained and several samples were taken
and subsequently analysed for major ions.

In this paper the results of a detailed examination of the measurements are presented and it is shown that the application of simple phase and chemical equilibrium data enables the limits of the possible origins of the various ions to be deduced. Specifically the origin of the species SO_4^{2-}, Cl^-, NH_4^+ and NO_3^- is considered, leading to an estimate of the relative contributions of the above anions to the observed acidity of the cloud water.

2. RESULTS AND DISCUSSION

Material is incorporated into cloud water either by dissolution of gases or as aerosol particles. In principle, aerosol could be incorporated either at the nucleation stage of cloud formation, or subsequently by either Brownian attachment, or by diffusophoresis. Brownian attachment is only important for Aitken nuclei, $r < 0.1$ μm, (Mason, 1971) whilst Goldsmith et al. (1963) deduced that diffusophoresis in a water vapour flux typically accounts for less than 1 per cent of material scavenged during the condensation cycle in the atmosphere. Thus uptake by gaseous diffusion and nucleation by particles are probably the main processes involved in scavenging by clouds. The cloud water drops eventually evaporate to form an aerosol of mixed composition which then undergoes several 'cloud cycles'. It follows that, unless significant chemical transformation takes place in cloud water droplets on a time scale comparable to a droplet growth and evaporation cycle, the measured cloud water composition should be related to the gas-aerosol composition of the associated neighbouring air mass. It is on this basis that the following discussion is developed.

2.1 Sulphate

The sulphate determined analytically in the laboratory could be derived from H_2SO_4 aerosol, $(NH_4)_2SO_4$ aerosol or from dissolved sulphur dioxide gas which oxidized to sulphate during collection and storage. Aerosols of intermediate stoichiometry exist (Tang, 1976), but can be considered to be a mixture of appropriate proportions of H_2SO_4 and $(NH_4)_2SO_4$.

Fig. 1 shows the composition of cloud water collected during a plume traverse on 28 January 1981. Six cloud water samples were obtained in this period and the mean concentrations were 2017 μeqs/l SO_4^{2-}, 803 μeqs/l NO_3^-, 1859 μeqs/l Cl^- and 2053 μeqs/l NH_4^+. These concentrations are much higher than those normally observed in precipitation. Fig. 1 shows the observed pH record and the SO_2 concentration profile. From the latter measurements and the observed cloud liquid water content, the concentration of dissolved SO_2 in equilibrium with pure water can be calculated as well as the expected equilibrium pH. The latter quantity is also shown on Fig. 1. Clearly the measured pH is much lower than could be accounted for simply by SO_2 dissolution.

The results from 3 other flights also show the observed pH values are lower than the calculated pH for all concentrations of SO_2 greater than ~10 ppb, implying that such a situation is normal.

The concentration of aqueous phase sulphur species that would be in equilibrium with 200 ppb SO_2 for aqueous solutions of different initial pH show that the predominant ion in each case is HSO_3^-. Even if sulphite species were all oxidized to SO_4^{2-} after collection the simple dissolutionn of SO_2 could clearly not account for the observed concentrations of SO_4^{2-} in the cloud water. Moreover, the pH of the stored sample is always comparable to that observed at collection and this suggests a negligible concentration of bisulphate at the time of collection, because oxidation or loss as SO_2 would have led to a change in pH.

It is thus clear that dissolved SO_2 cannot account for the observed pH of the cloud samples and that the measured sulphate levels do not arise simply from dissolved SO_2 which had oxidized during storage. The sulphate observed in the cloud water must therefore arise either from the oxidation of SO_2 within the droplet or from the capture of aerosol sulphate.

On the basis of laboratory experiments oxidation of SO_2 in cloud or rain droplets is possible through reaction with dissolved oxygen, ozone or hydrogen peroxide (Billingsley, et al., 1976, Penkett, et al., 1977). Using such data it is possible to calculate the expected progress of oxidation for any set of initial conditions of pH, SO_2 concentration etc. For O_2 and O_3 the reaction rate decreases with pH and thus the reactions are self-inhibiting but the reaction with H_2O_2 is insensitive to pH change (Cocks, McElroy and Wallis, 1982) and is thus the only possible route through which SO_2 oxidation could have been proceeding at the low observed pH levels shown in Fig. 1.

Using the data of Penkett et al. (1977) the progress of oxidation has been calculated for 200 ppb of SO_2. For significant oxidation to occur within the nominal time scale of a cloud droplet growth cycle an aqueous phase concentration of H_2O_2 of ~0.1 ppb would be required. Concentrations of H_2O_2 > 1.0 ppb have been reported in rural air, (Kelly and Stedman, 1979), but only in the presence of >40 ppb of ozone. During the present experiments only very low levels of O_3 were observed within the plume, <1 ppb, and photochemical activity was low. Consequently the H_2O_2 concentration would be expected to be <0.1 ppb (Cocks, 1981).

There seems no doubt, therefore, that aqueous phase oxidation could not have been proceeding at a significant rate at the time of observations because the pH was too low for reaction with O_2 or O_3 and because the oxidant level was probably too low for oxidation by H_2O_2. It follows from the above that the sulphate in cloud water must have derived from a combination of both background aerosol and aerosol formed earlier in the plume history.

Taking the observed concentration of sulphate in the cloud water along with the measured cloud liquid water content (0.62 g/m^3) and assuming no vertical concentration profile, it is possible to derive a value for the expected sulphate aerosol concentration below the cloud. The mean sulphate concentration of 2017 μ eq./litre gives a value of 58 $\mu g/m^3$ for the aerosol concentration. This compares with the normal ambient range of sulphate aerosol measurements of 1-25 $\mu g/m^3$ for a 24 hour average, (Harrison and Pio, 1981) and is thus quite credible for a plume 150 km downwind of the source.

On the 28.1.81 the background sulphate was ~10 $\mu g/m^3$. The sulphate formed in the plume in ~4 hours was therefore ~48 $\mu g/m^3$. This is equivalent to ~11.5 ppb SO_2 or an average transformation rate of ~1.4% per hour.

No a priori estimate of the relative proportions of H_2SO_4 and $(NH_4)_2SO_4$ in the aerosol is possible on the basis of this discussion. However, this will be deduced after the relative concentrations of other species have been defined below.

2.2 Chloride

The data for the chloride ion concentrations appropriate to Fig. 1 have been corrected for the contribution from sea salt using the measured Na and the known ratio of Na^+/Cl^- in sea salt (0.56). The possible sources are HCl gas and NH_4Cl aerosol. HCl is a primary pollutant and NH_4Cl would be a secondary product. The contribution from other chlorides should be small, the correction for sea salt already having been made.

HCl is a very soluble gas so that the partition between gas and aqueous phase is almost completely in the aqueous phase for atmospheric concentrations. For example, the concentration of HCl in equilibrium with the observed mean excess chloride concentration of 1859 µeq/l would be only ~3 × 10^{-4} ppb.

If the excess chloride in cloud derived from the gas phase HCl, and assuming no vertical structure, the concentration below the cloud would have to be ~25 ppb. The observed SO_2 concentration and the emission ratio of HCl and SO_2, lead to an expected value of 15-30 ppb so that the calculated value of 25 ppb would only be feasible if there had been no significant loss of HCl from the plume by wet or dry deposition.

Alternatively, if all the chloride were derived from NH_4Cl, then a concentration of ~40 µg/m^3 below the cloud would be required. This is too high to be realistic compared with normal aerosol measurements of excess chloride which are usually <2.5 µg/m^3 (e.g. Pio, 1981). Allowing for the standard deviation in such measurements, and the fact that they are 24 hour averages and not specifically in a plume a credible estimate of aerosol chloride would be 2.5 to 8 µg/m^3.

It seems reasonable to conclude that probably the bulk of excess chloride derived from gaseous HCl but there could be a contribution from particulate NH_4Cl in the range of 6 to 20%.

2.3 Ammonium

The species of interest are NH_3 gas, $(NH_4)_2SO_4$ aerosol, NH_4Cl aerosol and NH_4NO_3 aerosol. Ammonia gas from land sources can be considered the primary species and the aerosols secondary products.

Ammonia is very soluble in acid cloud drops and the gas phase concentration in equilibrium with the mean observed pH and concentration of NH_4^+ of 2053 µeq/ℓ is ~1 × 10^{-3} ppb. Ambient NH_3 measurements over land are usually in the range 0.1-2 ppb and are probably lower over the sea. As in the case of HCl, with such a low equilibrium gas phase concentration it is apparent that the observed concentration in cloud water could have derived from the gas phase. However, applying the same approach as above, the implied gas phase concentration outside the cloud would be ~28 ppb. This is far in excess of the ambient measurements <2 ppb (Healy and Pilbeam, 1974).

If all the ammonium derived from aerosols, $(NH_4)_2SO_4$, NH_4NO_3 and NH_4Cl, the implied concentration below the cloud would be ~22 µg/m^3. This is about twice the highest 24 hour average reported outside plumes (Healy and Pilbeam, 1974; Harrison and Pio, 1981), but is consistent with the high sulphate levels discussed above. Furthermore ammonium aerosols exceed NH_3 gas in the ambient air by a factor of 5-10. Thus, it seems reasonable to conclude that the bulk of ammonium in the cloud water samples derives from aerosols with a possible contribution of about 3-7% from the gas phase if ambient NH_3 levels are appropriate.

2.4 Nitrate

The oxides and oxyacids of nitrogen form a complex series of equilibria in the gas and liquid phases involving NO, NO_2, N_2O_3, N_2O_4, N_2O_5, HNO_2 and HNO_3. The thermochemical parameters of some of these species are not known and estimates such as those of Schwartz and White (1981) must be used to calculate equilibrium concentrations.

The observed gas phase concentrations during the traverse relating to Fig. 1 were NO = 50 ppb and NO_2 = 12 ppb and the mean observed nitrate ion in cloud water was 803 µeq/ℓ. The nitrate ion is the state of lowest free energy and if equilibrium were achieved with the observed gas phase concentrations, a 2 molar solution of nitrate would be expected. Clearly, equilibrium had not been achieved in this case. The concentration of nitrite in equilibrium with the observed gas phase composition would be ~0.35 μmoles/ℓ, ~25% dissociated. (Nitrous acid is much weaker than sulphurous acid.) Thus even if equilibrium between the gas phase and nitrous acid is achieved the nitrite contribution to the observed nitrogen in solution must be very small even if it is oxidized to nitrate subsequent to collection, and before analysis.

The oxidation and decomposition of nitrous acid in solution has been the object of many studies (see Lee and Schwartz, 1981 and references therein). The reaction can proceed via two overall stoichiometries:

$$3HNO_2 \rightarrow HNO_3 + 2NO + H_2O \qquad \ldots (2)$$

$$2HNO_2 \rightarrow NO + NO_2 + H_2O \qquad \ldots (3)$$

Reaction (2) is favoured when the gas phase resistance to mass transfer is high. The empirical stoichiometry for NO_3^- formation is 0.13 → 0.33. An upper limit to the rate of nitrate production can be deduced by assuming that (a) nitrite is an equilibrium with the observed gas phase NO_x and (b) that only reaction (2) occurs. Using rate data from various alternative sources (Abel et al., 1928; Komyana et al., 1978; Lee and Schwartz, 1981) yields reaction rates in the range $7.10^{-13} - 1.4.10^{-12}$ m s^{-1} which are extremely low. Oxidants could increase the rate and thus Penkett (1972) and Bhattachacharyya et al. (1977) reported rate constants for oxidation by O_3 and H_2O_2 respectively which lead to rates of $1.1.10^{-11}$ m s^{-1} (O_3 = 50 ppb) and $1.4.10^{-10}$ m s^{-1} (H_2O_2 = 1 ppb). Clearly such rates are much too low to produce significant NO_3 in the ~4 hours of plume travel and although concentrations of NO_x are higher nearer the source, it appears that the observed nitrate is not formed from oxidation in solution and that nitrate in the cloud water must derive either from gas-phase HNO_3 or from nitrate aerosol.

In order to calculate the formation of HNO_3 in the gasphase a complex kinetic model is required. However a reasonable estimate may be made by noting that the key reaction is

$$NO_2 + OH \rightarrow HNO_3 \quad k = 8E-12 \ cm^3p^{-1}s^{-1} \qquad \ldots (1)$$

(Rodhe et al., 1981). Taking a value of 1 to 3 × 10^6 particles/cc for OH, the rate is equivalent to 0.3 to 1 ppb h^{-1} in the plume where the observations were made.

If it is assumed that gas phase HNO_3 is in equilibrium with the cloud water drops then a concentration of 7 × 10^{-4} ppb is required. Ambient measurements of gas phase HNO_3 are in the range 0.2 to 5 ppb (Kelly and Stedman, 1979), so with a low equilibrium concentration it is clear that the observed concentration in cloud water could have derived from gas phase HNO_3. The implied gas phase concentration below the cloud thus becomes 11 ppb which, although higher, is not unreasonable compared to the reported non-plume ambient measurements. It is also reasonable in terms of the amount that could be formed by gas phase reaction (1) integrated over the plume history. A concentration of 11 ppb represents a nominal transformation rate of NO_x to HNO_3 of ~3.8% per hour which again is consistent with predictions of detailed reactive plume models (Cocks, 1981).

The possibility that nitrate derives from NH_4NO_3 aerosol has also been examined. If all the observed nitrate in the cloud water came from aerosol, the implied concentration below the cloud would be 30 μg/m^3.

This is high compared with normal ambient measurements, which are in the range 1-10 $\mu g/m^3$, (Harrison and Pio, 1981). A concentration of up to ~20 $\mu g/m^3$ in the plume would, however, be credible. Therefore, up to two thirds of the nitrate could have derived from aerosol on this basis.

However, there is a further constraint on the concentration of NH_4NO_3 aerosol in the presence of sulphate if it is assumed that the aerosol is an 'internal' mixture of nitrate, ammonium and sulphate as opposed to pure ammonium nitrate. In such a multicomponent system, sulphate displaces nitrate and the vapour pressure of HNO_3 above such an aerosol is high. Tang (1980), has estimated the vapour pressures of such a system and although not all the thermochemical parameters are known it is clear that the vapour pressure of HNO_3 in equilibrium with such a mixed aerosol is strongly humidity dependent. Using Tang's results and the mean observed molar ratio of nitrate to sulphate in the cloud water, the vapour pressure of HNO_3 above the aerosol would be: 43 ppb at 95% RH, 197 ppb at 90% RH and 1790 ppb at 75% RH. Such aerosols would be 85%, 72% and 42% by weight water respectively. However, such calculations do not conserve total mass on nitrate in the two phases. If a cloud drop with the mean composition were to evaporate, then the products, conserving mass, would be, 5.6 ppb HNO_3 + 15 $\mu g/m^3$ nitrate aerosol at 98.5% RH, 10.8 ppb and 0.8 $\mu g/m^3$ at 90% RH and 11 ppb and 0.2 $\mu g/m^3$ at 60% RH. This could account for very low levels, (<3 $\mu g/m^3$), of nitrate aerosol found on the other flights (Marsh, Clark et al., 1980, 1981a, b and c and Glover, Lightman et al., 1980, 1981a and b). The partial pressure of NH_3 above such aerosols is almost independent of relative humidity and is ~0.02 ppb —much lower than ambient measurements and the aerosols should be stable with respect to NH_3. The proportion of the nitrate observed in cloud water which is derived from aerosol depends on the humidity.

The above estimate of the proportions of aerosol and gas assume that equilibrium is attained. The rate of attainment of these equilibrium is not known, but, if time scales of the order of tens of minutes are involved, then compositions will be much less sensitive to relative humidity in the shorter term, allowing NH_4NO_3 to exist as an aerosol well below cloud level.

2.5 Acidity

Table 1 summarizes the results from the above discussion. The seven species of interest are $H_2SO_4(s)$, $(NH_4)_2SO_4(s)$, $HCl(g)$, $NH_4Cl(s)$, $NH_3(g)$, $HNO_3(g)$ and $NH_4NO_3(s)$. If the concentrations of all these and the $H^+(\ell)$, $SO_4^{2-}(\ell)$, $NO_3^-(\ell)_3$, $Cl^-(\ell)$ and $NH_4^+(\ell)$ in cloud water are expressed in units of neqs/m^3, then the following equations can be written:

$$SO_{4(\ell)} = H_2SO_{4(s)} + (NH_4)_sSO_{4(s)} \qquad \qquad \cdots (4)$$

$$Cl_{(\ell)} = HCl_{(g)} + NH_4Cl_{(s)} \qquad \qquad \cdots (5)$$

$$HN_{4(\ell)} = NH_{3(g)} + NH_4Cl_{(s)} + NH_4NO_{3(s)} + (NH_4)_2SO_{4(s)} \qquad \cdots (6)$$

$$NO_{3(\ell)} = HNO_{3(g)} + NH_4NO_{3(s)} \qquad \qquad \cdots (7)$$

$$H_{(\ell)} = HNO_{3(g)} + HCl_{(g)} + H_2SO_{4(s)} \qquad \qquad \cdots (8)$$

where the LHS represents the cloud and the RHS represents the implied below cloud composition.

The discussions above enabled estimates to be made of all the species in the equations, expect $H_2SO_{4(s)}$ and $(NH_4)_{2(s)}$. Using equations (4) to (8), it is then possible to estimate $H_2SO_{4(s)}$.

Table 2 shows the matrix for the composition appropriate to 96% relative humidity. This yields a proportion of 21% $H_2SO_{4(s)}$ to 79% $(NH_4)_2SO_{4(s)}$. Further, with the relative contribution of nitrate gas and aerosol, the contribution of sulphur to the acidity of cloud water is 16%, that of $HNO_{3(g)}$ is 27% and that of $HCl_{(g)}$ is 57%. However, in view of the uncertainty associated with the nitrate, the relative contribution of sulphur is in the range 13 to 28%.

It is pertinent to question whether the concept of nitrate aerosol is very meaningful. At equilibrium the normal state of nitrate is gas phase HNO_3 and nitrate aerosol could only form in close proximity to clouds. This is consistent with the very low aerosol nitrate observed on several flights (Marsh, Clark et al., 1980, 1981a, b and c and Glover, Lightman et al., 1980, 1981a and b). It could be argued, therefore, that the lowest figure for the contribution of sulphur to cloud water acidity is the most appropriate one.

The meteorological situation on 28.1.81 was rather unusual and the contribution of HCl would normally be expected to be lower at this range (105 km), and consequently the sulphur contribution would normally be higher. The observations are consistent with the simple concept that the sequence of loss from the plume and hence the range of influence on the composition of cloud and precipitation is chloride, nitrate and then sulphate. On the 28.1.81 this process was slow and extended to a greater range than normal. Such a sequence leads to an increase in the ratio SO_4/NO_3 with distance (Rodhe et al., 1981).

3. CONCLUSIONS

A period of cloud sampling on the aircraft flight of 28.1.81 has been examined in an attempt to account for the observed concentrations of SO_4^{2-}, NO_3^-, Cl^- and NH_4^+ and the pH of the cloud water.

1. It is conncluded that the simple dissolution of SO_2 cannot account for either the observed pH or the concentration of sulphate in the cloud water.

2. No oxidation of SO_2 to SO_4 was occurring in the cloud drops at the time of observation because of lack of oxidant.

3. The sulphate concentration in the cloud water is accounted for by sulphate aerosol with a concentration below the cloud of 58 $\mu g/m^3$.

4. The chloride concentration in the cloud water is accounted for by dissolved HCl gas and a <20% contribution from NH_4Cl aerosol.

5. The observed nitrate in the cloud water cannot be accounted for simply in terms of dissolved NO or NO_2 and no oxidation of nitrite to nitrate was occurring in cloud droplets.

6. The nitrate in cloud water can be accounted for in terms of a mixture of dissolved HNO_3 gas and HN_4NO_3 aerosol. The proportion of the mixture is very sensitive to RH at high relative humidity, but below ~94% RH the gas contribution completely dominates.

7. From the deduced contributions of HCl gas and HNO_3 gas to the cloud drops it is possible to calculate the relative contributions of H_2SO_4 and $(NH_4)_2SO_4$ aerosol to the composition of the cloud drop. This is in the range 17.0% H_2SO_4 at 60% RH to 36% at 98.5% RH. The uncertainty depends on the magnitude of the dissolved HNO_3 gas.

8.   The pH of the cloud water is accounted for by 57% dissolved HCl gas, 10% dissolved HNO_3 gas and 13% from H_2SO_4 aerosol.
9.   In terms of the pattern of precipitation expected of downwind of major sources, the observations are consistent with the wet deposition sequence of Cl^-, NO_3^- and SO_4^{2-} but because of the unusual meteorological conditions, for this particular flight the travel distance of this sequence was greatly extended.

4.   REFERENCES

Abel, E. and Schmid, H., 1928, Z. Phys. Chem., _132_, 55, _134_, 279 and _136_, 135

Bhattacharyya, P.L. and Veeraraghavan, R., 1977, 'Reaction between Nitrous Acid and Hydrogen Peroxide in Perchloric Acid Medium', Int. Journal Chem. Phs. _IX_, p. 629-640

Billingsley, J., Kallend, A.S. and Marsh, A.R.W., 1976, 'Washout of Sulphur Dioxide by Rain', CERL Note No. RD/L/N 85/76

Cocks, A.T., 1981, Private Communication

Cocks, A.T., McElroy, W.J. and Wallis, P.G., 1982, 'The Oxidation of Sodium Sulphite Solutions by Hydrogen Peroxide', CERL Note No. RD/L/221N81

Glover, G.M. and Lightman, P., 1980, 'CERL/EPRI Programme on the Fate of Atmospheric Emissions, Background Flight of 17.6.80', CERL Memorandum No. LM/CHEM/237

Glover, G.M., Hamilton, P.M. and Lightman, P., 1981a, 'CERL/EPRI Programme on the Fate of Atmospheric Emissions, Background Fligt of 21.10.80', CERL Memorandum No. LM/CHEM/2

Glover, G.M., Gatford, C., Lightman, P. and Webb, A.H., 1981b, "CERL/EPRI Programme on the Fate of Atmospheric Emissions, Instrumental Comparison Flight of 11.12.80', CERL Memorandum No. LM/CHEM 286

Goldsmith, P., Delafield, H.J. and Cox, L.C., 1963, 'The Role of Diffusiophoresis in the Scavenging of Radioactive Particles from the Atmosphere', Q.J. Roy. Met. Soc., _89_, p. 43-51

Harrisonn, R.M. and Pio, C.A., 1981, 'Apparatus for Simultaneous Size-Differentiated Sampling of Optical and Sub-optical Aerosols: Application to Analysis of Nitrates and Sulphates', JAPCA _31_, No. 7 p. 784-787

Healy, T.V. and Pilbeam, A., 1972, 'Ammonia and Related Atmospheric Pollutants at Harwell', AERE Report No. R 6231

Hegg, D.A. and Hobbs, P.V., 1981, 'Cloud Water Chemistry and the Production of Sulphates in Clouds', Atmos. Envir., _15_, p. 1597-1604

Kelly, T.J. and Stedman, D.H., 1979, 'Measurements of H_2O_2 and HNO_3 in Rural Air', Geophysical Research Letters, _6_, No. 5, p. 375-378

Komiyama, H. and Inoue, H., 1978, 'Reaction and Transport of Nitrogen Oxides in Nitrous Acid Solutions', J. Chem. Eng. Japan, _11_, No. 1, p. 25-32

Lee, Y.N. and Schwartz, S.E., 1981, 'Reaction Kinetics of Nitrogen Dioxide with Liquid Water at Low Partial Pressure', J. Phys. Chem., _85_, p. 840-848

Marsh, A.R., Clark, P.A., Webb, A.H. and Fisher, B.E.A., 1980, 'CERL/EPRI Programme on the Fate of Atmospheric Emissions, Hercules Flight of 18 June 1980', CERL Memorandum No. LM/CHEM241

Marsh, A.R., Clark, P.A., Laird, C.K., Moore, D.J. and Wallis, R., 1981b, "CERL/EPRI Programme on the Fate of Atmospheric Emissions, Hercules Flight of 29 January 1981', CERL Memorandum No. LM/CHEM/298

Marsh, A.R., Clark, P.A., Dear, D.J., Fisher, B.E.A., Moore, D.J. and Webb, A.H., 1981c, 'CERL/EPRI Programme on the Fate of Atmospheric Emissions, Hercules Flight of 22 October 1980', CERL Memorandum No. LM/CHEM/296

Mason, B.J., 1971, 'The Physics of Clouds', Oxford University Press, 2nd
   Edition
OECD, 1974-1976, Long Range Transport of Air Pollutants, Data is issued by
   the Norwegian Institute for Air Research and given in a series of reports
   LRTAP 4/74, 4/75, 18/75, 19/75, 20/75 and 2/76
Okita, T., Kaneda, K., Yanaka, T. and Sugai, R., 1974, 'Determination of
   Gaseous and Particulate Chloride and Fluoride in the Atmosphere', Atmos.
   Envir., 8, p. 927-936
Penkett, S.A., 1972, 'Oxidation of SO_2 and Other Atmospheric Gases by Ozone
   in Aqueous Solution', Nature, 240, p. 105-106
Penkett, S.A., Jones, B.M.R. and Brice, K.A., 1977, 'Rate of Oxidation of
   Sodium Sulphite Solutions by Oxygen, Ozone and Hydrogen Perioxide and its
   Relevance to the Formation of Sulphate in Cloud and Rainwater', AERE
   Report No. R 8584
Pio, C.A., 1981, Ph.D. Thesis, University of Lancaster
Rodhe, H., Crutzen, P. and Vanderpol, A., 1981, 'Formation of Sulphuric and
   Nitric Acid in the Atmosphere During Long-range Transport', Tellus, 33,
   p. 132-141
Schwart, S.E. and White, W.H., 1980, 'Solubility Equilibria of the Nitrogen
   Oxides and Oxyacids in Dilute Aqueous Solution', In: 'Adv. Environ. Sci.
   Eng., Vol. 4, J.R. Pfafflin and E.N. Ziegler, Eds., Gordon and Breach,
   New York, In press
Skartveit, A., 1982, 'Wetscavenging of Sea Salts and Acid Compounds in a
   Rainy, Coastal Area', Atmos. Envir. in press
Tang, I.N., 1976, 'Phase Transformation and Growth of Aerosol Particles
   Composed of Mixed Salts', J. Aerosol Science, 7, p. 361-371
Tang, I.N., 1980, 'On the Equilibrium Partial Pressures of Nitric Acid and
   Ammonia in the Atmosphere', Atmos. Envir., 14, p. 819-828
Walters, P.T., Moore, M.J. and Webb, A.H., 1982, 'A Separator for obtaining
   samples of cloud water in aircraft', Atmos. Envir. in press

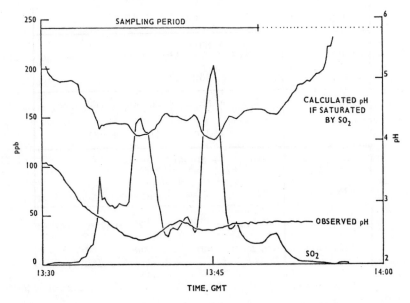

Fig. 1.   Cloud Sampling on 28/1/81

Table 1: Cloud, Gas and Aerosol Concentrations

Species	Cloud water $\mu eq/\ell$	Gas concentration in equilibrium with cloud drop ppb	Implied below cloud concentration expressed as	
			gas ppb	aerosol $\mu g/m^3$
SO_4	2017	65000	13	58
Cl	1859	2.5E-4	25	40
NH_4	2053	1.0E-3	28	22
NO_3	803	7.1E-4	11	30

Table 2: Composition Below Cloud at RH = 96%

Concentrations in neqs/m^3

$SO_{4(\ell)}$ =	$H_2SO_{4(s)}$ + $(NH_4)_2SO_{4(s)}$		
1250	268	981	
$Cl_{(\ell)}$ =	$HCl_{(g)}$ + $NH_4Cl_{(s)}$		
1152	922	230	
$NH_{4(\ell)}$ =	$NH_{3(g)}$ + NH_4Cl + NH_4NO_3 + $(NH_4)_2SO_4$		
1273	0 + 230 + 61 + 981		
$NO_{3(\ell)}$ =	$HNO_{3(g)}$ + $NH_4NO_{3(s)}$		
498	437 61		
$H_{(\ell)}$ =	$HCl_{(g)}$ + $HNO_{3(g)}$ + $H_2SO_{4(s)}$		
1627	922 437 268		

PRELIMINARY MEASUREMENTS FROM AN AIRCRAFT INTO
THE CHEMICAL COMPOSITION OF CLOUDS

F.G. RÖMER, J.W. VILJEER, L. VAN DEN BELD, H.J. SLANGEWAL,
and A.A. VELDKAMP
N.V. KEMA, Environmental Research Department.
Arnhem, the Netherlands
H.F.R. REIJNDERS
National Institute of Public Health, Laboratory for
Environmental Chemistry, Bilthoven, the Netherlands

Summary

Using an aircraft preliminary investigations have been
carried out into the chemical composition of clouds.
Rather high concentrations of H_3O^+, Cl^-, NO_3^-, SO_4^{2-}, NH_4^+
and H_2O_2 were measured. Except H_2O_2 these components are
scavenged very fast during transport over short distances
over source regions. Moreover H_3O^+, NO_3^- and SO_4^{2-} are
likely to be formed by chemical reactions of gaseous precursors with oxidants in the liquid phase. These reactions
may cause the strong decrease in high background H_2O_2 concentrations (ppm range).
Occasionally the chloride concentration was found to be
much higher than may be expected from rain network data.

1 INTRODUCTION

    Until now the usual rainwater measurements have been carried
out at ground level stations mainly structured within a measuring
network and also by means of event sampling (1, 2). In studying
the relation between emissions of SO_2 and NO_x from specific
source regions and the resulting acidity of rain collected on
ground level, network data are of limited value due to long
sampling times. From combined event sampling and meteorological
studies the composition of rainwater can sometimes be related
to the originating air masses or be classified more generally
(2).
    As we initially were interested in the acid producing com-
ponents and precursors (SO_2 and NO_x), the analysis included
the determination of sulphate, nitrate, ammonium, hydronium ion,
chloride, pH and hydrogen peroxide.
    At the moment little is known on the European continent
about the relations between the emission of acid forming components as SO_2 and NO_x and the composition of cloudwater and
about the relation between composition of cloudwater and of
rain on ground level.
    A suitable way to gather information on these problems is
to collect cloud and rainwater together with gaseous and particulate components using an aircraft. Provided the measurements
can be combined with reliable trajectory analyses ultimately
this research can lead to measuring strategies which enable
the determination of a source receptor relation and of budget
calculations as estimates of trans-frontier transport of air

pollution. We decided to start with pliot investigations within clouds enabling us to answer questions related to the operation of the sampling devices under several conditions.

2 INSTRUMENTATION

The aircraft used is a Piper Navajo Chieftain flown by Geosens B.V., Rotterdam.

A description of aircraft and instruments is given elsewhere (3). The sampling devices for cloud and rainwater have been provided by Atmospheric Sciences Research Center, State University, New York (Albany) (4).

The chemical analysis used are described in references 5-8.

Absolute volumes needed for the separate analysis are in the range of 50 to 400 µl. To run the analysis automatically, a total volume of 5 ml is needed for the single determination of anions, ammonium, hydronium ions and hydrogen peroxide.

3 RESULTS

The cloud and rainwater contents of clouds were determined during several flights and in different seasons. Some characteristic volumes (mean water quantities sampled along the track) are presented in table 1.

Table 1 Liquid water content of clouds

date	amount of water (mg/m³) cloud	rain	temp. (°C)	altitude (m)	cloud type (m)	cloud* cover	location
81-11-19	30- 70(4)[+]	20	3-5	1000-1300	strato-cumulus (800-1700)	6/8	Belgium (Ardennes)
81-11-30	70-120(4)	130;150	0-4	500	strato-cumulus; cumulo-nimbus (400-1700)	7/8	Netherlands (Gouda-Scheve-ningen v.v.)
82-03-18	20- 65(4)	35; 90	2,5-5,5	450	stratus; cumulus (400-1700)	7/8	Netherlands
82-07-06	40-550 (16)	20-365 (16)	8-12,5	1200-1400	strato-culumus (700-1800)	7/8	Netherlands

[+] () number of samples

As the minimum sample volume needed for a complete analysis
for the components under consideration, amounts to 5 ml, the
maximum spatial resolution for cloudwater is 50 s or about 4 km
and for rainwater 25 s or about 2 km (speed 70 m/s, impaction
surface for cloud and rain samplers being 25,45 cm² and 49,75 cm²
respectively). Because of the inhomogeneity of water density
distribution in clouds over a flight track, the minimum resolution cannot be determined reliably.

Typical results for chloride, sulphate, nitrate, hydronium
ion and pH are given in table 2 for a flight on 82-07-06.

Table 2   Results from analysis of cloud and rainwater collected
          over the Netherlands during a flight on 82-07-06

| | concentration range* (µMol/l) | |
	cloud	rain
nitrate	8 - 860	10 - 1460
sulphate	18 - 400	14 - 535
chloride	16 - 475	1 - 695
hydronium	5 - 745	5 - 745
ammonium	12 - 1055	13 - 915
pH	3,2- 5,5	3,1 - 6,1

*
16 samples for cloud and for rainwater; the concentrations in this
publication are not corrected for sea spray

The values given are only meant as indications of concentrations which can occur.

It was found that the maximum concentrations of nitrate,
sulphate, hydronium and ammonium ions are 10 to 20 times as high
as the monthly mean concentrations determined in rainwater samples for corresponding substances from the Dutch measuring network (3).

In table 3 an overview of the measured H_2O_2 concentrations
is given.

It is noticed that high concentrations of hydrogen peroxide
can occur in the liquid phase. These high concentrations can be
explained by the large Henry's law constant for H_2O_2 and occur
in background or hardly polluted air masses from the North Sea.
Hydrogen peroxide concentrations measured above the continent
and on ground level indicate that its concentration is strongly
reduced because of reactions with precursors (NO_x and SO_2). As
not much data have been published, we have also summarized some
results of the H_2O_2 concentrations from samples (snow, rain) collected at ground level. Though these concentrations are rather
low, they can vary to a large extent.

During the measurement on 82-07-06, the changes over the
tracks found for the rainwater content, concentration levels of
H_3O^+ and SO_4^{2-} are given in the figures 1-3 respectively. These
profiles are comparable with concentration profiles for Cl^-, NH_4^+
and NO_3^-. For this reason the profiles for Cl^-, NH_4^+ and NO_3^- are
not given. Moreover the profile for concentrations in cloudwater
samples generally show a comparable picture. Considering the
concentration profiles, it is noticed that the concentration
level increases going from west to east along the transport di-

Table 3 Results of the determinations of hydrogen peroxide in cloud and rainwater collected during several flights and in background air on ground level*

date	tracks/locations	altitude (m)	H_2O_2 concentration (µMol/1) mean cloud	mean rain	range cloud	range rain	remarks
82-08-09	Rotterdam (NL) – Stockholm (S)	2000-3000	25 (9)+	23 (9)	9-59	0,5-59	pH 3,9-5,2 SO_2 < 4 ppbv westerly wind; outward flight: strato-cumulus; return flight: cumulus; stratocumulus
82-07-27	Hull (GB) : Den Helder (NL)	150-2000 150-2000	38 (2) 88 (1)	41 (2)	32-44	20-62	stratocumulus spirals northerly wind SO_2 < 2 ppbv; NO_2 < 3 ppbv; NO < 1 ppbv; O_3 40-45 ppbv
82-07-06	see Fig. 1	1100-1400	17 (7)	17(16)	<1,2 -45	<1,2 -71	measurements in background and in polluted air
81-12-04 81-11-30	Flevopolder (NL)	ground level		0,05(60)		0,01-0,2	measurements in background air during plume washout experiments
81-12-29	Riezlern (A)	ground level at 1200 m above sea level		0,15 (2)		0,05-0,25 snow	

+ () number of determinations
* incident measured maximum values during summer 1981 at Arnhem: 2 µMol/1

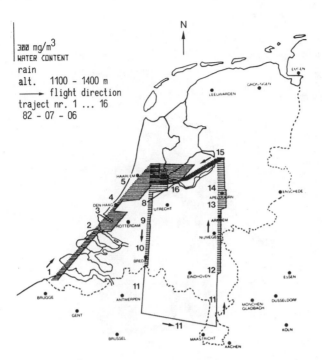

Fig. 1 Flightpattern and rainwater contents

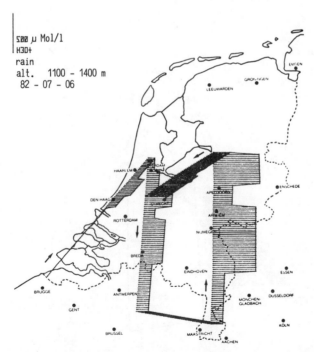

Fig. 2 Concentration profile for hydronium ions

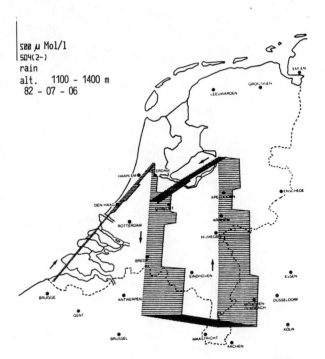

Fig. 3 Concentration profile for sulphate

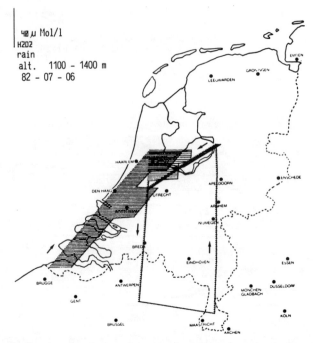

Fig. 4 Concentration profile for hydrogen peroxide

rection. Probably this may be explained by the effect of the western industrial/urban region in the Netherlands and possibly Belgium (track 11).

The concentration profile for H_2O_2 is given in figure 4. Comparing figure 4 and figures 2 and 3, it can be established that with the increase of the concentration of several components the concentration of H_2O_2 decreases. Therefore, an important role of hydrogen peroxide in the liquid phase formation of acid is proposed. This proposition is supported by the finding that hydrogen peroxide concentration decreases at a high rate with increasing sulphate concentration. This effect is graphically presented in figure 5.

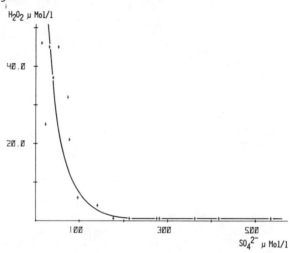

Fig. 5 Diagram for hydrogen peroxide and sulphate concentration in rainwater measured over the Netherlands during a flight on 82-07-06

In figure 6 the NO_3^-/SO_4^{2-} ratios and nitrate concentrations along the flightpattern are given. From this figure the importance of nitrate ion is illustrated. During transport of clouds over rather short distances (some tens of km), the NO_3^-/SO_4^{2-} ratio can strongly change by scavenging gaseous and particulate nitrogen and sulphur species. However, the contribution of NO_x and SO_2 to acidity must be studied further.

High chloride concentration profiles can occur over the country. These high concentrations (maximum values 695 and 475 µMol/l in cloud and rainwater respectively) cannot be explained from sea spray. Thus it can be expected that a large part of chloride originates from anthropogenic sources and that chloride plays also an important role in acidification. Since this is not in agreement with expected theories, further study into this phenomenon has to be done.

4 GENERAL REMARKS

As we were able to sample both cloud and rainwater (from a precipitating cloud (4), we measured on a few occasions a complete different sampling yield for both cloud and rainwater samplers.

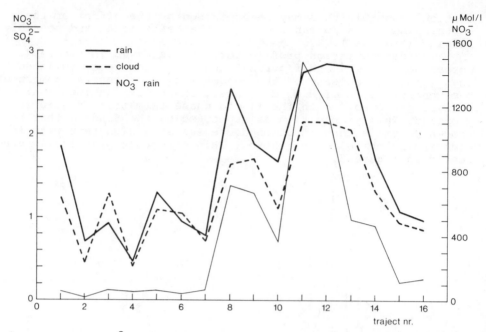

Fig. 6 NO_3^-/SO_4^{2-} molar ratios and NO_3^- concentrations for 16 rainwater samples (flight 82-07-06)

Probably this depends on meteorological conditions (cloud type, temperature, droplet size distribution). If size radius of the droplets is too small, only small volumes of cloudwater can be collected. This must be taken into account when planning flight schemes for collecting cloud and rainwater.

The inhomogeneity of clouds and the droplet size distribution are factors that can influence the measurements in an unpredictable way and therefore continuous monitoring of cloud liquid water content should be pursued.

During a few experiments remarkably high concentrations of the main components (Cl^-, NO_3^-, SO_4^{2-}, H_3O^+ and NH_4^+) were measured. On the contrary, high concentrations of hydrogen peroxide were measured in background air. Further, the concentration of H_2O_2 seems to be inversely proportional to those of other components.

5 REFERENCES

1 Ridder, T.B., Reijnders, H.F.R., Van Esseveld, F., Wegmans, R.C.C. and Mook, W.G. (1981) Chemical precipitation over the Netherlands. Survey 1978-1980, RIV-KNMI report; KNMI 156-3a, RIV 217810-004
2 Slanina, J. Römer, F.G., Asman, W.A.H. (1983) Investigation of the source regions for acid deposition in the Netherlands, Workshop on Acid deposition, Berlin (GFR)
3 Viljeer, J.W. and Van Duuren, H. (1977) The measurements by aeroplane of atmospheric pollution, in particular in smoke plumes (in Dutch); Elektrotechniek 55, 540-547
Römer, F.G., Van Duuren, H., Elshout, A.J. and Viljeer, J.W.

(1979) Messungen aus Flugzeugen: Verbreitung und Umwandlung primärer luftfremder Komponenten in Rauchfahnen; VGB Konferenz Kraftwerk und Umwelt 1979, 134-140 (Essen, GFR)

4   Mohnen, V.A. (1980) Cloudwater collection from aircraft, Atmospheric Technilogy $\underline{12}$

5   Slanina, J. et al., personal communication

6   Technicon Industrial Systems (1977) Individual/simultaneous determination of nitrogen and/or phosphorus in BD acid digests, Ind. Methods 229-74 W/b, Tarry Town, New York

7   Small, H., Stevens, T.S. and Bauman, W.C. (1975) Novel ion exchange chromatographic method using conductimetric detection, Analyt. Chem. $\underline{47}$, 1801-1809

8   Kok, G.L., Holler, T.P., Lopez, M.B., Nachtrieb, H.A. and Yuan, M. (1980) Chemiluminescent method for determination of hydrogen peroxide in ambient atmosphere, Envir. Sci. Technol. $\underline{12}$, 1972-1976

MODEL CALCULATIONS OF THE EFFECT OF SO_2, NO_x OR HC
EMISSION CHANGES ON REGIONAL SCALE SULPHATE AND
NITRATE PRODUCTION

Ø. Hov
Norwegian Institute for Air Research
P.O.Box 130, N-2001 Lillestrøm, Norway

Summary

A comprehensive model of the gas phase chemistry of hydrocarbons, nitrogen oxides and sulphur dioxide is applied
to assess the effect of emission changes of SO_2, NO_x and
HC on the production of secondary species like ozone,
sulphate and nitrate on a regional spatial scale and a
temporal scale of a few days. Transport of air from the
UK is taken as a case study. A reduction in UK NO_x
emissions only may cause increased sulphate production in
air transported towards Scandinavia. A reduction in UK
SO_2 emissions may result in a similar reduction in the
sulphate concentrations, but depending on the evaluation
of the uncertainties in the underlying chemical process,
it is calculated that a 50% SO_2 emission control in the
UK may result in only 30-35% sulphate concentration reduction in an ageing air mass moving over the sea from
the UK.

1. INTRODUCTION

The chemical turnover of SO_2, NO_x and HC in the atmosphere
is strongly linked together. Change in the emission of one
species or group of species may affect the degradation and
turnover of other species in non-linear ways, due to the inter-
dependencies among the various groups of species. An increase
in the emission of oxides of nitrogen may bring about a reduction of pH in cloud droplets, slowing down the oxidation of
sulphur dioxide to sulphate. An increase in NO_x may also
depress the formation of hydroxyl radicals, which would cause
a drop in the sulphate formation. These mechanisms would
influence the partition of airborne sulphur into sulphur dioxide and sulphate. Consequently, the deposition pattern of
sulphur would be altered, because the atmospheric residence
times of SO_2 and sulphate are different.
    The question of how changes in the emission of SO_2, NO_x or
HC, or a combination of them all, affect the deposition of
SO_2, sulphate and HNO_3 on a spatial scale of the order of
1000 km and a temporal scale of a few days, will be discussed
below.

2. MODEL DESCRIPTION

The model applied here is the one developed and described by Derwent and Hov (1979, 1980, 1982). An illustration of the model is given in Fig. 1. It is a box type model with a mixing height of 1300 m, which is assumed to be stationary over the UK for nearly 3 days. The polluted air where primary and secondary pollutants have accumulated is then followed in a Lagrangian sense over a sea surface for nearly 5 days e.g. towards Scandinavia. Fair weather is assumed. Gas chemical conversion processes take place, together with dry deposition and heterogeneous loss of species like H_2O_2, N_2O_5 and CH_3O_2H, with a first order decay coefficient of $1x10^{-5}s^{-1}$ (characteristic time $\simeq$ 28 h).

40 species are emitted (NO, NO_2 SO_2, CO and 36 different hydrocarbons). The contribution from eight source categories (petrol engine exhaust, diesel engine exhaust, motor vehicle evaporative losses, stationary combustion, solvent usage, industrial processes, petroleum industry and natural gas leakage) is estimated, and a summary is given in Table 1. Approximately 145 intermediate species and end products and 300 reactions, 28 of them photolytic are required to describe the turnover of the precursors. The chemical reactions involved are described in detail by Derwent and Hov (1979, 1980), and revised according to the kinetic data reviews published by Atkinson et al. (1979, 1982).

Each chemical species satisfies the continuity equation in the form

$$\frac{Dc}{dt} = P_{ch} + \frac{F_c}{H} - (L_{ch} + L_{het} + \frac{v_d}{H}) \, c$$

where Dc/dt denotes the total time derivative of the concentration c, P_{ch} and L_{ch} c photochemical production and loss terms, L_{het} c parameterized heterogeneous loss, F_c emission flux of the compound in question and v_d the deposition velocity, specified in Table 2.

The formation of sulphuric acid is initiated through the reaction

(R1) $OH + SO_2 \rightarrow HSO_3$ $k_1 = 1.1x10^{-12}$ (Calvert et al., 1978)

The rate coefficient given applies to 300 K temperature. The pressure dependent expression for k_1 given e.g. by NASA (1981), gives a value of $1.2x10^{-12}$ for k_1 at 285K. If the low and high pressure limits are increased by one standard deviation, to give an indication of what is currently believed as an upper limit estimate for k_1, a value of $1.9x10^{-12}$ is found for 300K, $2x10^{-12}$ at 285K. The significance of the choice of k_1 was tested in the model calculations.

The fate of the HSO_3 radical is not well known. The following mechanism is included in the model:

$$HSO_3 + O_2 \rightarrow HSO_5 \qquad\qquad 1.0x10^{-15} \text{ (Cox, 1974)}$$
$$HSO_5 + NO \rightarrow HSO_4 + NO_2 \qquad 6x10^{-12}$$
$$HSO_4 + O_2 \rightarrow .. \rightarrow \alpha_1\, HO_2 + H_2SO_4 \quad 1x10^{-15}$$

The unit of all rate coefficients is $cm^3 (molecule \times s)^{-1}$ unless otherwise stated. Reaction R1 is the rate determining step. α_1 is a factor between zero and one. The choice of this parameter is significant because with $\alpha_1 < 1$, the oxidation of sulphur dioxide through R1 is a sink for hydroxyl radicals while if $\alpha_1 = 1$, the sulphur chemistry does not affect the odd hydrogen radical chemistry. In the model calculations reported here, $\alpha_1 = 1$ unless otherwise stated.

The formation of nitric acid takes place through the reaction

$$\text{(R2)} \quad NO_2 + OH \xrightarrow{k_2} HNO_3 \qquad\qquad k_2 = 1.1 \times 10^{-11}$$

(Hampson and Garvin, 1978)

The coupled set of differential equations expressing the conservation of mass for each species, is solved numerically using a qssa-method shown to be accurate to within ±1% (Derwent and Hov, 1979).

3. MODEL RESULTS

The model was run for nearly three days with constant UK emissions mixed over a depth of 1300 m, cpr. Fig. 1. Afterwards the emissions were cut off. The chemical development in the air mass which then was assumed to travel over the ocean, was followed. In Fig. 2 is shown the development of ozone, where a maximum of approx. 135 ppb was reached after more than two days over the sea. The hydrocarbon and NO_x precursors were then sufficiently depleted for the loss process to balance the production and later on the concentration of ozone declined very slowly. Clearly, the chemical life time of ozone in ageing air masses is very long (one week or more, cpr. Hov et al., 1978). In the figure is also shown the accumulated emissions of NO_x together with the amount of ozone removed at the ground. The total amount of ozone generated was 155 ppb or 7 times the integrated NO_x emissions.

In Fig. 3 is shown the relative distribution (by volume) with time of the concentrations of sulphate, airborne SO_2 and SO_2 removed at the sea surface. The concentration of SO_2 was depleted, while most of the sulphur was deposited as SO_2. In Fig. 4 is shown a similar graph of the distribution of nitrogen containing species. Only a minor amount of nitrogen containing aerosol was formed, due to the slow rate coefficient chosen for the heterogeneous loss ($L_{het} = 1 \times 10^{-5} s^{-1}$), and nitric acid was assumed to remain in the gas phase or removed by dry deposition. In Fig. 5 is shown the relative distribution of gaseous nitrogen species and it is quite clear that the primary emitted compounds are very rapidly depleted.

In Table 3 is shown maximum concentrations on the days 3 and 7 for a number of species and compared with measured values in the UK. Nitric acid was overestimated since the computed concentration represented the sum of nitric acid gas and aerosol.

3.1 Emission scenarios

The effect of altered emissions on the formation of ozone,

sulphate and nitric acid as the air masses aged over the ocean, was investigated. The scenarios chosen are listed in Table 4.

The development of the concentrations of O_3, SO_2, sulphate, deposited SO_2, HNO_3, deposited HNO_3, NO_2 in the air and deposited, PAN in the air and deposited, and nitrogen containing aerosols, in the model run shown in Figures 2-5 over the sea surface, is shown in Figures 6-8. In the same figures the influence of the various emission scenarios is shown relative to the 1975 UK emission situation. Some conclusions can be drawn from the figures:

1. With respect to ozone, changes in the level of NO_x emissions by ± 50% tend to have little effect after a few days over a sea surface. Increase in the HC level by 50% (with or without simultaneous NO_x emission increase), seems to lead to a 20% higher ozone concentration after 2 days or more. SO_2 control affects the concentration of ozone insignificantly.

2. For SO_2, only SO_2 control has a significant impact. Changes in the emissions of HC or NO_x or both by ±50%, tend to reduce the SO_2 oxidation and leave more SO_2 in the air over a sea surface (10-15%). SO_2 control by 50% reduces the SO_2 concentration by 50% as well. This demonstrates that the concentration of hydroxyl is only insignificantly influenced by SO_2, as long as the odd hydrogen radical concentration is not influenced by the sulphur chemistry.

3. The amount of dry deposited SO_2 follows the pattern for SO_2.

4. With respect to sulphate aerosol, an increase in HC or decrease in NO_x by 50% seem to increase the sulphate level slightly (a few per cent). A 50% increase in NO_x emissions or both HC & NO_x, is calculated to suppress the sulphate concentration over the sea compared with the present situation. Again, a 50% SO_2 control results in a 50% sulphate reduction.

5. For airborne and deposited HNO_3 over the sea surface, NO_x or HC & NO_x increase by 50% result in 50% increase or more on the 7th day, compared to the present situation. SO_2 control or 50% HC increase have only a slight impact, while NO_x control by 50% seems to reduce the amount of airborne and deposited HNO_3 by nearly 50%.

3.2 Formulation of the sulphate production

It was indicated above that there are many uncertainties in the formulation of how SO_2 is converted to sulphate through the reaction with hydroxyl, and how fast the process is.

The influence of the choice of k_1, α_1 and emission flux of SO_2 on the concentration of ozone 6h after the emissions were abolished (i.e. at noon on day 3), is shown in Table 5. There is a very significant difference between the value 60 ppb in case 4, with no feedback to the hydrogen radical chemistry ($\alpha_1 = 0$), the fast k_1 rate coefficient (2×10^{-12}) and uncontrolled SO_2 emissions (f=1), to the case with $\alpha_1 = 1$, $k_1 = 2 \times 10^{-12}$ and f = 1 where the ozone mixing ratio was 98 ppb. In the corresponding cases with the recommended values for k_1

(1.1×10^{-12}), 69 and 94 ppb of ozone were computed. It is concluded that with UK 1975 emissions of SO_2, if HO_2 is not re-
generated through the SO_2 to sulphate conversion process, the
intensity of the SO_2 emissions significantly influences the
turnover also of NO_x and the hydrocarbons. In this case, a
reduction in SO_2 enhances the chemical activity since the
hydroxyl radical sink through R1 becomes less efficient.
In such a situation, control of SO_2 is counter productive with
respect to ozone formation (Table 5), and the concentration of
sulphate remains at 60-70% of the level with uncontrolled SO_2
(Fig. 9).
      The suppression of the hydroxyl radical concentration
which is calculated to take place when $\alpha_1 = 0$ compared to the
case with $\alpha_1 = 1$ in reaction R1, increases the time scales
for sulphate and nitric acid production, and increases the
chemical lifetimes of SO_2 and NO_2. The ratio between the dry
removal of SO_2 and the wet removal of sulphate will increase,
a similar situation will occur for NO_2 and nitric acid.

ACKNOWLEDGEMENT

      In this paper is reported preliminary results from a
project about the turnover of NO_x and SO_2 in the lower atmosphere, funded by the Norwegian Department of Environment. A
steering group is set up, consisting of Anton Eliassen (The
Norwegian Meteorological Institute), Ivar Isaksen (University
of Oslo), Harald Dovland and Øystein Hov (NILU) and the project
assistant, Jon Jerre.

REFERENCES

Apling A.J., Sullivan E.J., Williams M.L., Ball D.J., Bernard
      R.E., Derwent R.G., Eggleton A.E.J., Hampton L. and Waller
      R.E. (1977) Ozone concentrations in south-east England
      during the summer of 1976. *Nature* 269, 569-573.
Atkins D.H.F., Cox R.A. and Eggleton A.E.J. (1972) Photo-
      chemical ozone and sulphuric acid aerosol formation in the
      atmosphere over southern England. *Nature* 235, 372-376.
Atkinson R., Darnall K.R., Lloyd A.C., Winer A.M. and Pitts
      J.N. Jr. (1979) Kinetics and mechanisms of the reaction of
      the hydroxyl radical with organic compounds in the gas
      phase. *Adv. Photochem.* 11, 375-488. John Wiley.
Atkinson R., Lloyd A.C. and Winges L. (1982) An updated
      chemical mechanism for hydrocarbon/NO_x/SO_2 photooxidations
      suitable for inclusion in atmospheric simulation models.
      Atmospheric Environment 16, 1341-1355.
Calvert J.G., Su F., Bottenheim J.W. and Strausz O.P. (1978)
      Mechanism of the homogeneous oxidation of sulphur dioxide
      in the troposphere. *Atmospheric Environment* 12, 197-226.
Campbell M.J., Sheppard J.C., and Au B.F. (1979) Measure-
      ment of hydroxyl concentration in boundary layer air by
      monitoring CO oxidation. *Geophys. Res. Lett.* 6, 175-178.
Cox R.A. (1974) The photolysis of nitrous acid in the presence
      of carbon monoxide and sulphur dioxide. *J. Photochem.* 3,
      291-304.

Cox R.A., Derwent R.G. and Sandalls F.J. (1976) Some air
   pollution measurements made at Harwell, Oxfordshire
   during 1973-1975. AERE-R 8324, H.M. Stationery Office.
Davis D.D., Heaps W., Philen D. and McGee T. (1979) Boundary
   layer measurements of the OH radical in the vicinity of an
   isolated power plant plume: SO_2 and NO_2 chemical conversion
   times. *Atmospheric Environment* 13, 1197-1203.
Derwent R.G. and Hov Ø. (1979) Computer modelling studies of
   photochemical air pollution formation in North West Europe.
   AERE R-9434, Her Majesty's Stationery Office, London.
Derwent R.G. and Hov Ø. (1980) Computer modelling studies of
   the impact of vehicle exhaust emission controls on photochemical air pollution formation in the United Kingdom.
   Environ. Sci. & Technol. 14, 1360-1366.
Derwent R.G. and Hov Ø. (1982) The potential for secondary
   pollutant formation in the atmospheric boundary layer in a
   high pressure situation over England. *Atmospheric Environment* 16, 655-665.
Eliassen A., Hov Ø., Isaksen I.S.A., Saltbones J. and Stordal
   F. (1982) A Lagrangian long-range transport model with
   atmospheric boundary layer chemistry. *J. Appl. Meteor.*
   (to appear in the November issue).
Hampson R.F. and Garvin D. (1978) Reaction rate and photochemical data for atmospheric chemistry 1977. NBS Special
   Publication 513, Washington, D.C.
Hov Ø., Hesstvedt E. and Isaksen I.S.A. (1978) Long-range
   transport of tropospheric ozone. *Nature* 273, 341-344.
NASA (1981) Chemical kinetic and photochemical data for use in
   stratospheric modelling. Evaluation Number 4: NASA panel
   for data evaluation. NASA, Jet Propulsion Laboratory,
   California Institute of Technology, Pasadena JPL 81-3.
Penkett S.A., Sandalls F.J. and Lovelock J.E. (1975) Observa-
   tions of peroxyacetyl nitrate (PAN) in air in southern
   England. *Atmospheric Environment* 9, 139-140.
Penkett S.A., Sandalls F.J. and Jones B.M. (1977) PAN measure-
   ments in England - Analytical methods and results. *VDI
   Berichte* 270, VDI-Verlag GmbH, Düsseldorf pp. 47-54.
Perner D., Ehhalt D.H., Pätz H.W., Platt U., Röth E.P. and
   Volz A. (1976) OH-radicals in the lower troposphere.
   Geophys. Res. Lett. 3, 466-468.

Table 1: Average U.K. emissions in molecules $cm^{-2}s^{-1}$ (Derwent and Hov 1979).

Species	Emission	Species	Emission
NO	2.81×10^{11}	CH_3CHO	4.94×10^8
SO_2	4.23×10^{11}	C_2H_5CHO	3.70×10^8
CO	2.94×10^{12}	C_3H_7CHO	1.51×10^8
CH_4	1.22×10^{12}	iC_3H_7CHO	6.79×10^7
C_2H_6	4.07×10^{10}	C_4H_9CHO	3.46×10^7
C_3H_8	8.64×10^9	CH_3COCH_3	3.48×10^9
nC_4H_{10}	2.19×10^9	$CH_3COC_2H_5$	2.43×10^9
iC_4H_{10}	9.88×10^9	methylpropylketone	1.24×10^7
nC_5H_{12}	1.80×10^{10}	methyl-i-propylketone	1.24×10^7
iC_5H_{12}	3.20×10^{10}	CH_3OH	1.24×10^7
C_2H_4	2.56×10^{10}	C_2H_5OH	1.75×10^{10}
C_3H_6	9.58×10^9	1-butene	3.30×10^9
C_2H_2	2.41×10^{10}	2-butene	4.43×10^9
toluene	1.81×10^{10}	2-pentene	2.02×10^9
o-xylene	7.04×10^9	1-pentene	3.62×10^9
m-xylene	7.04×10^9	2-methyl-1-butene	3.12×10^9
p-xylene	7.04×10^9	3-methyl-1-butene	2.19×10^9
ethylbenzene	7.04×10^9	2-methyl-2-butene	5.04×10^9
HCHO	2.84×10^9	butylene	1.24×10^9
		benzaldhyde	3.87×10^8

† Base year 1975.

Table 2: Deposition velocity v_d (cms^{-1})[a]

Species	Land surface		Sea surface	
	day	night	day	night
O_3	0.6	0.06	0	0
HNO_3	0.8	0.8	0.8	0.8
SO_2	0.8	0.8	0.8	0.8
NO_2	0.5	0.1	0	0
PAN	0.2	0.02	0	0

a) For further discussion and references, see Eliassen et al. (1982).

Table 3: Comparison of model calculated concentrations with observations in rural southern England (in ppbv, aerosols in $\mu g\ m^{-3}$).

Species	Calculated maximum concentration 3d day	7th day	Measured	
OH	2.4×10^{-4}	1.2×10^{-4}	$> 2.8\times10^{-4}$	(Perner et al., 1976)
			$1.2\times10^{-5}-1.6\times10^{-4}$	(Campbell et al., 1979)
			$1.3\times10^{-4}-4.3\times10^{-4}$	(Davis et al., 1979)
HO_2	0.01	0.03		
SO_2	12	1.0	7.5	CDS*
nC_4H_{10}	0.8	0.3	2.3	CDS
C_2H_4	0.3	0.001	2.3	CDS
HCHO	2.8	0.8	2.7	CDS
CH_3CHO	1.0	0.3		
NO_2	4.9	0.4	5.6	CDS
NO	1.5	0.03	2.1	CDS
PAN	1.8	1.0	2.0-8.9	(Penkett et al., 1977)
PPN	0.3	0.1	1/6 of PAN+	
HNO_3	7.6	3.2	1.7	CDS
O_3	69	130	<250	(Apling et al., 1977)
Sulphate aerosol‡	18	35	10-70	(Atkins et al., 1972)

* CDS: Cox et al. (1976)
+ The ratio between gas chromatogram peaks of PPN and PAN is fairly constant in all measurements and is usually around 1:6-1:8 (Penkett et al., 1975)
‡ In $\mu g\ m^{-3}$

Table 4: Emission scenarios

Species Case	NO_x	HC	SO_2
Present (1975)	1	1	1
(a) HC increase	1	1.5	1
(b) HC&NO_x increase	1.5	1.5	1
(c) NO_x increase	1.5	1	1
(d) NO_x control	0.5	1	1
(e) SO_2 control	1	1	0.5

Table 5: Mixing ratio for O_3 (ppb) 6h after the emissions are abolished (noon, day 3) in the case of 1975 UK emissions, for various choices of parameters for the reaction

$$(R1) \quad SO_2 + OH \xrightarrow{k_1} \ldots \to \alpha_1 HO_2 + H_2 SO_4$$ and choices of emission control for SO_2 ($F_{SO_2} \times f$).

Case	k_1 a)	α_1	f	O_3
1	1.1×10^{-12}	1	1	94
2	1.1×10^{-12}	0	1	69
3	1.1×10^{-12}	0	0.5	78
4	2×10^{-12}	0	1	60
5	2×10^{-12}	0	0.5	72
6	2×10^{-12}	1	1	98
7	2×10^{-12}	1	0.5	93

a) in cm^3 (molecule x s)$^{-1}$

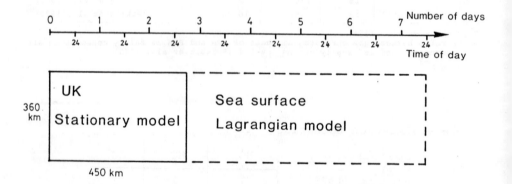

Figure 1: Illustration of the model. An air mass is modelled by locating a stationary box of dimensions 360x450 km^2 and depth 1300 m over the UK for nearly 3 days, whereafter the air mass is followed in a Lagrangian sense for more then 4 days over the sea.

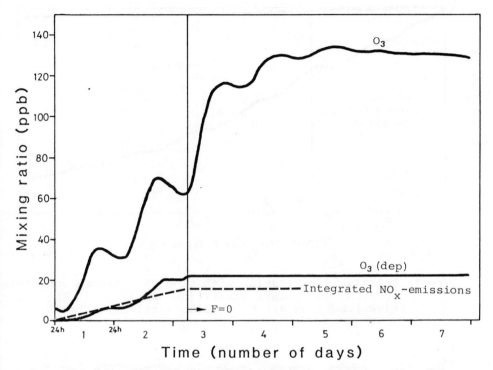

Figure 2: Development of gas phase and deposited ozone with time. No emissions after day 3 (indicated by the vertical arrow). The integrated NO_x emissions are also shown.

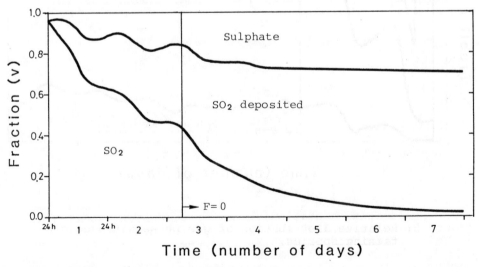

Figure 3: Relative distribution of the sulphur budget with time (on a volume basis): SO_2 (gas phase), deposition and aerosol.

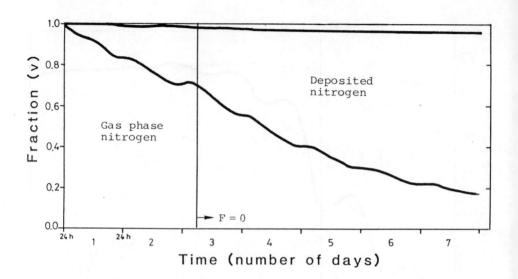

Figure 4: Relative distribution with time of nitrogen containing species: gas phase, deposition and aerosol. The relative fractions are on a volume (molecule) basis.

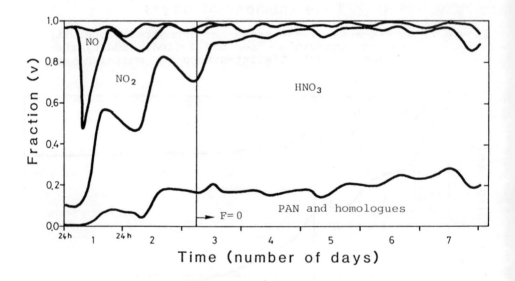

Figure 5: Relative distribution of gas phase nitrogen containing species.

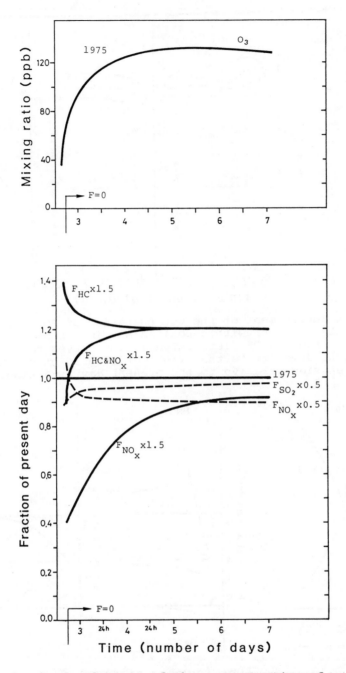

Figure 6: The development of the concentration of ozone over sea from the UK (upper curve), together with the effect of altered emission scenarios relative to the present day situation (lower curves).

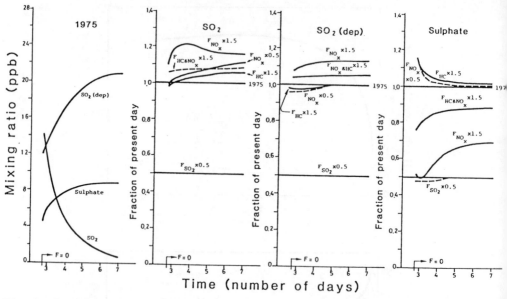

Figure 7: The development of the concentration of SO₂ and sulphate together with deposited SO₂ in air passing over
sea from the UK (left curves). The curves to the
right demonstrate the effect of altered emission
scenarios relative to the present day situation.

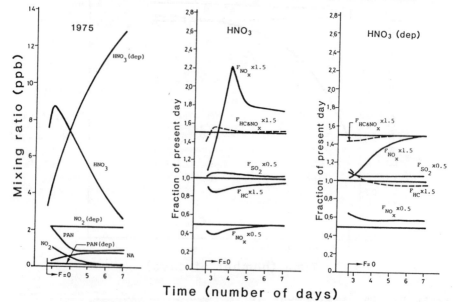

Figure 8: The development of the concentration of HNO₃, NO₂,
PAN, nitrogen containing aerosols, together with deposited PAN, HNO₃ and NO₂ (left curves). The curves
to the right demonstrate the effect of altered emission scenarios relative to the present day situation.

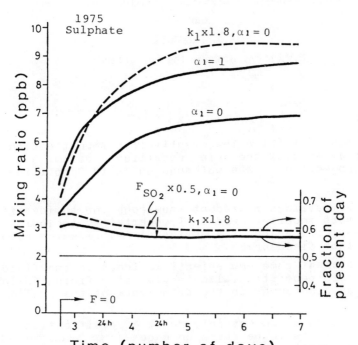

Figure 9: The effect of altering the chemical parameters
describing the sulphate production through the reac-
tion $SO_2 + OH \rightarrow \cdot_{[} \cdot \rightarrow \alpha_1 HO_2 + H_2SO_4$, $k_1 = 1.1 \times 10^{-12}$ cm^3
(molecule x s)$^{-1}$. The emissions are kept at the UK
1975 level.

WORKSHOP CONCLUSIONS

by

S. BEILKE

Umweltbundesamt, Pilotstation Frankfurt, W.-Germany

and

A.J. ELSHOUT

N.V. KEMA, Arnhem, The Netherlands

The workshop dealt with the quantitative investigation of
different aspects of the acid deposition problem.
The main objective of the workshop on "Acid Deposition" was
to

- summarize our present knowledge and to identify the
  most important areas of uncertainty including a
  specification of a series of key research needs in
  this field

- present some new scientific ideas in the field of
  acid deposition, also in view of defining following-
  up programmes in the frame-work of the COST-actions.

The major areas of uncertainty and the specification of key
research needs is to be found on pages 16 to 22 in these proceedings. Requisite research outlined in these proceedings
has mainly resulted from discussions within the Acid Deposition Task Force of project COST 61a bis.
Some use was also made of three workshops(Alta,USA,1978;
La Jolla,USA,1982;Santa Monica,USA,1982) published in the
corresponding workshop proceedings.

Furthermore,a series of research needs was specified on
effects(Ulrich,1982)and on the formation of atmospheric aci-
dity(Cox and Penkett,1982) in these proceedings.

In the following section some of the most important results
and conclusions of this workshop will be summarized. Some use
was also made of a series of papers presented at the 4th
International Conference on Precipitation Scavenging,Dry
Deposition and Resuspension(Santa Monica,California,USA,
29 November - 3 December 1982).

The deposition of acidic substances to the earth surface may
proceed in different ways. Acids are contained in rain-and
fogdroplets,in snowflakes and aerosol particles and reach the
ecosystem via these bulk H^+-carriers.
Acids like gaseous nitric acid(HNO_3) or gaseous hydrochloric
acid(HCl) can be deposited directly or acids may be formed

if gases like SO_2 and NO_x are deposited and react with water. All of these different mechanisms may contribute to the acidification of ecosystems and are collectively called acid deposition.

The workshop in Berlin mainly dealt with one facet of the acid deposition issue, the problem of acid rain.

The papers presented have shown that of the large number of factors which are thought to be of importance for the acid deposition problem, only relatively few are well known. On the one side it is recognized that we have environmental effects on aquatic and terrestrial ecosystems as well as on materials in various European regions attributed at least in part to acid deposition. It is furthermore recognized that we have acid deposition in wide areas of Europe originating from air pollution mainly of anthropogeneous origin. But if we go one step further and try to construct a well established link between these two separately well recognized facts, our knowledge rapidly becomes less complete.

As far as the atmospheric facet of the acid deposition problem is concerned, this begins with a definition and determination of atmospheric acidity.

The most extensive approach so far to determine atmospheric acidity was presented by Brosset(1983) using the approach of thermodynamic phase equilibria within an atmospheric system consisting of a liquid and gaseous phase containing the main elements of the acidity-determining martix. As far as the acidity within the liquid phase is concerned, Brosset has stressed that a pH-value alone provides a rather incomplete information on the acid properties of rainwater and that measurements have to focus on those parameters which determine the pH and not only on the pH-value itself.

Another important question is whether, and to what extent, total acid deposition has increased since the onset of industrialisation and especially during the last 3 decades in Europe. In the light of the substantial increase of man-made emissions of SO_2 and NO_x in north-west Europe since 1860, it is reasonable to assume that total deposition of acidic substances in the industrialized regions has also considerably increased since that time. A quantification of the increase of total deposition of acidic substances on the basis of the available measurements can, however, not be given. Trend measurements for dry acidic deposition are virtually non-existent and trend developments of total acidity of rainwater have never been attempted in spite of a considerable data base generated during the last two decades. Instead of total acidity of rainwater which is a measure for total acidic wet deposition, a series of trend developments for the free acidity(pH values) of rainwater were reported in the literature. Three papers presented during the Berlin-workshop and published in these proceedings have shown that previous publications indicating an increase of the geographical extent of the regions with low rainwater pH and a general decrease of rainwater pH in central Europe during the last two decades should be taken with caution(Kallend,1983;Winkler,1983; Ridder and Frantzen,1983).

Kallend has shown that of the 120 sites of the European Atmospheric Chemistry Network with 5 or more years data 29 reveal a statistically significant trend of decreasing annual average precipitation pH during the period 1956-1976.Five stations showed an opposite trend.

Similiar conclusions were drawn by Winkler(1983) on the basis of a critical investigation of literature release data.He concluded that since the late 30th the precipitation pH has not changed very much in central Europe in spite of a doubling of SO_2-emissions since that time.

According to Winkler,at a pH of ca. 4.2,a further increase of acidic gases such as SO_2 does not necessarily result in a further pH-decrease but rather to a spreading out of the area of acid rain.

No conclusive picture concerning pH-trends emerged from the paper of Ridder and Frantzen(1983) entitled" Acid Precipitation over the Netherlands".There seem to be some indications that yearly precipitation pH averages at two sites in the Netherlands have decreased between 1956 and the mid sixtith and increased after that time.

The rainwater acidity in Europe is to a large extent determined by the acidic gases SO_2 and NO_x.The major portion of sulfate and nitrate in rainwater can be traced back to these precursor gases.An important problem in this context is the question of what the sulfate and nitrate concentrations in rainwater are and what are their trends.

According to Wallen(1980),both sulfate and nitrate concentrations in rainwater in central Europe are much higher than the corresponding values in the neighbor regions.

As far as sulfate-trends in precipitation in Europe are concerned,the overall picture is not uniform.

Based on measurements of 25 stations of the Background Air Pollution Monitoring Network,Wallen(1980) calculated mean sulfate concentrations in rainwater for the period 1972-1976 and compared these values with corresponding values for the late 50th published by de Bary and Junge(1963).

A comparision reveals for Europe as a whole a mean increase by 50% with considerable differences on a regional scale.

When comparing these two mean sulfate concentrations one should keep in mind that sampling and analytical procedures were not the same in the 1950's and 1970's.

According to Kallend(1983),of the 120 stations of the European Atmospheric Chemistry Network with 5 or more years data 23 stations showed a statistically significant increase of sulfate and 1 station a decrease between 1956 and 1976.

In contrast to sulfate,an increase of nitrate in rainwater seems to be documented.Of the 120 stations investigated by Kallend,55 showed an increase of nitrate in rainwater and none a decrease between 1956-1976.

The mean increase rate was 6%/year.

Another possible mean for investigating trends are analyses
of deep ice cores in the arctic and antarctic regions.
In contrast to rainwater analyses, such deep ice core analyses
allow under certain circumstances an evaluation of total acidic
deposition trends. As most man-made emissions of SO_2 and NO_x
occur in the industrialized regions of the northern hemisphere
(for instance according to Cullis and Hirschler(1980), over
90% of the man-made sulfur is emitted in the northern hemisphere) and as these gases and especially their oxidation products can be transported over hundreds to thousands of kilometers, one should expect to find an increase of sulfate and
nitrate in polar ice in the northern hemisphere from the onset
of industrialisation if the emissions of SO_2 and NO_x have considerably increased since that time.

In fact, Herron(1980) found for the time period before 1900
until today a mean increase of sulfate deposition by a factor
of 2 to 2.5, for nitrate a factor of 2 whereas for chloride
no such increase could be observed.
Natural and man-made contributions to today's sulfate and
nitrate values are about equal.

In contrast to the polar regions of Greenland, deep ice core
measurements in Antarctica did not reveal an increase of
sulfate and nitrate deposition during the last 100 years.

Another aspect of the acid deposition problem concerned the
investigation of natural free "background acidity".
The paper of Delmas and Gravenhorst(1983) and publications of
Charlson and Rodhe(1982) have shown that the natural background acidity is very sensitive to traces of atmospheric
pollutants such as sulfur compounds resulting in a wide range
of rainwater pH's originating from naturally emitted substances. Thus a rainwater pH of 5.6 cannot be used as a reference
value for an unpolluted background atmosphere.
For instance, rainwater pH's on the small island of Amsterdam
(37^oS, 77° E , 55 km^2, Indian Ocean) which is not influenced
by anthropogeneous pollution are between 3.5 and 5.5
(Charlson, 1982).
On the other hand, in large areas of India and China pH values
between 6 and 8 are observed which are most likely not associated with man-made emissions.

Another important aspect of the acid deposition issue concerns
the formation of atmospheric acidity. Although a considerable
knowledge exists on this matter, there are still very important
areas of uncertainty some of which are specified in the review
paper of Cox and Penkett(1983).

Thus we will confine ourselves to a few words about the formation of sulfuric acid in atmospheric droplets by presenting
a brief historical look to place the present situation in
context and to emphazise the important results achieved under
project COST 61a.

The fact that "acid rain" is a phenomenon that occurs in
various European regions has been known for more than 120
years.
Measurements of the chemical composition of rainwater began

in the middle of the last century(Lawers et al.,1861).
In 1872 the first British Chief Alkali Inspector R.Smith
published a book entitled" Air and Rain - The Beginnings of
a Chemical Climatology".Based on a series of measurements in
different regions of the United Kingdom and of Germany,he
concluded that sulfuric acid plays an important role for
acidification of rainwater and that ammonia acts as a neutralizing agent.
At the turn of the century,the German chemists Bigelow(1898)
and Titoff(1903) measured that metal catalysts such as manganese,iron and copper are responsible for the oxidation of
sulfur(IV) in aqueous solutions.
For more than 50 years this mechanism was thought to be the
only process by which sulfate and sulfuric acid is formed in
atmospheric droplets.
Then came the important work of Junge and Ryan(1958) suggesting that ammonia is responsible for sulfate formation in
atmospheric droplets.This process was believed to be the
dominant mechanism for more than 15 years until 1974.
In 1974 it could be shown within project COST 61a that ammonia
only plays a minor role for SO_2-oxidation and that other
mechanisms had to be found which are effective in the low
pH-range normally encountered in atmospheric droplets
(Beilke and Barrie,1974;Barrie,Beilke and Georgii,1974).
Such a mechanism was suggested by the important work of
Penkett,Brice and Eggleton(1975) again within project COST
61a.On the basis of laboratory experiments these authors
suggested that strongly oxidizing agents such as ozone and
hydrogen peroxide(H_2O_2) provide the dominant mechanism for
sulfate formation in atmospheric droplets.
At the end of COST 61a in 1976 the question of what the importance of the H_2O_2-reaction is under real atmospheric conditions could not be answered due to a lack of H_2O_2-concen-
tration measurements both in the gaseous-and droplet phase.
Some recently presented papers seem to confirm that sufficiently high H_2O_2-concentrations are formed in cloud droplets
which may lead to a rapid oxidation of sulfur(IV) to sulfur(VI)
(Römer et al.,1983;Chameides and Davis,1982;Schwartz et al.,
1982).
For stratiform and/or stratocumulus clouds Römer et al.(1983)
and Schwartz et al.(1982) found that H_2O_2 and sulfate in cloud
water were inversely correlated suggesting that sulfur(IV)
is rapidly oxidized by H_2O_2 resulting in a non-coexistence of
these two gases.
On the other hand,Richards et al.(1982) found that sulfur(IV)
and H_2O_2 may well coexist if sufficient formaldehyde is pre-
sent.Formaldehyde was found by them in all cloud water samples,
and is known to react with sulfur(IV) to form hydroxymethanesulfonic acid(HMSA).According to Richards et al.(1982),this
adduct contributes to the high sulfur(IV) concentrations they
observed in cloud water and to an inhibition of the reaction
between H_2O_2 and sulfur(IV).As HMSA is a strong acid,the
authors concluded that SO_2 and formaldehyde may produce acidity in cloud water without an oxidation of the sulfur(IV).

An important point in this context is to measure the chemical
composition of cloud water in various types of clouds inclu-

ding precipitating and non-precipitating clouds,fog,haze ect..
Two workshop papers dealt with measurements of the chemical
composition of different clouds(Römer et al.,1983;Marsh,1983).
Römer et al.(1983) measured the chemical of precipitating
stratocumulus and cumulonimbus clouds over the Netherlands
and Belgium and found rather high concentrations of H^+,NH_4^+,
Cl^-,NO_3^-,$SO_4^=$,and H_2O_2.The authors concluded that,except for
H_2O_2,these compounds were incorporated into cloud droplets
very soon after the clouds had passed over known source regions.
Marsh(1983) studied the acidity and chemical composition of
mostly stratiform clouds over the North Sea and found much
higher concentrations in clouds than in precipitation for all
ions.He concluded that no oxidation of SO_2 to sulfate was
occurring in the cloud droplets at the time of observation
i.e.sulfate in cloud water was accounted for by incorporation
of sulfate aerosol.

In the USA and in Canada a series of cloud-and fog acidity
measurements were carried out which were recently presented
at the 4th International Conference on Precipitation Scavenging,Dry Deposition and Resuspension(Santa Monica,USA,
29 November - 3 December 1982)
(Daum et al.,1982;Schwartz et al.,1982;Radke,1982;Hegg and
Hobbs,1982;Leatch et al.,1982;Waldman et al.,1982).
These papers contain some new information about the chemical
composition of cloud-and fogwater including some results on
interstitial concentrations of most compounds which are relevant to the formation of acidity in cloud-and fog droplets.
The papers will be published in 1983 in the corresponding
conference proceedings.

Summarizing the outcome of the acid deposition workshop in
Berlin,the papers presented represent an up-to-date state of
the art survey in the field of acid deposition at least in
the countries of the European Communities.The most important
areas of uncertainty could be indicated and a series of research needs specified which should provide a sound basis for
defining the possible COST 61a follow-up programme(1984-1985)
in which more emphasis will be placed on investigating
different aspects of the acid deposition problem.

References:

    Barrie,L;Beilke,S.and Georgii,H.W.(1974)
        SO2-removal by cloud-and raindrops as affected by
        ammonia and heavy metals.Proceedings:Precipitation
        Scavenging Symposium,Champaign,Illinois,USA,
        October 14-18,1974.

    Beilke,S.and Barrie,L.(1974)
        On the role of NH_3 in heterogeneous SO_2-oxidation in
        the atmosphere.Paper presented at Ispra(Italy),
        October 1974.Technical Symposium COST 61a.

    Bigelow,S.L.(1898)
        Zeitschrift für Phys.Chemie,26,493.

Brosset,C.(1983)
    Characterisation of acidity in natural waters.
    Paper presented at Workshop on "Acid Deposition",
    Berlin,Reichstag, 9 September,1982.This volume.

Charlson,R.J.and Rodhe,H.(1982)
    Factors controlling the acidity of natural rainwater.
    Nature,Vol.295,No.5851,pp.683-685.

Charlson,R.J.(1982)
    Background acidity.Paper presented at MPI Mainz,
    October 1982.

Cox,R.A.and Penkett,S.A.(1983)
    Formation of atmospheric acidity.Paper presented at
    Workshop on "Acid Deposition",Berlin,Reichstag,
    9 September,1982. This volume.

Cullis and Hirschler(1980)
        cited in:Herron,M.M.(1982)
    Impurity Sources of F^-,Cl^-,NO_3^- and $SO_4^=$ in Greenland
    and Antarctic Precipitation.Journal Geophysical
    Research,Vol.87,No.C4,April 20,1982,p.3053.

Chameides,W.L.and Davis,D.D.(1982)
    The coupled gas-phase/aqueous -phase free radical
    chemistry of a cloud.Paper presented at SCADDER
    conference,Santa Monica,USA, 29 Nov.-3 Dec.1982.

Daum,P.H.;Schwartz,S.E.and Newman,L.(1982)
    Studies of the gas and aqueous phase composition of
    stratiform clouds.Paper presented at SCADDER
    conference,Santa Monica,USA, 29 Nov.- 3 Dec.1982.

Delmas,R.and Gravenhorst,G.(1983)
    Background precipitation acidity.Paper presented at
    Workshop on "Acid Deposition",Berlin,Reichstag,
    9 September 1982. This volume.

Hegg,D.A.and Hobbs,P.V.(1982)
    The relative importance of particulate and gas
    scavenging to the inclusion of sulfate and nitrate
    in cloud water.Paper presented at SCADDER conference,
    Santa Monica,USA, 29 Nov.-3 Dec.1982.

Herron,M.M.(1982)
    Impurity Sources of F^-,Cl^-,NO_3^- and $SO_4^=$ in Greenland
    and Antarctic Precipitation.Journal Geophys.Research,
    Vol.87,No.C4,April 20,1982.

Junge,C.E.and Ryan,T.G.(1958)
    Study of the SO_2-oxidation in solution and its role
    in atmospheric chemistry.Q.J.R.met.Soc.84,pp.46-55.

Kallend,A.S.(1983)
    Trends in the acidity of rain in Europe:a re-exami-
    nation of European atmospheric chemistry network
    data.Paper presented at Workshop on "Acid Deposition",
    Berlin,Reichstag, 9 September 1982. This volume.

Lawers,Gilbert and Pugh(1861)
    Paper cited in:Acidity of Rainfall in the UK-a preli-

liminary report.UK review group on acid rain,prepared
for the UK Department of the Environment,June 1982.

Leaitch,W.R.;Strapp,J.W.;Wiebe,H.A.,and Isaac,G.A.(1982)
Measurements of scavenging and transformation of
aerosol inside cumulus.Paper presented at SCADDER
conference,Santa Monica,USA, 29 Nov.- 3 December 1982.

Marsh,A.R.W.(1983)
Studies of the acidity and chemical composition of
clouds.Paper presented at Workshop on "Acid Deposition",Berlin,Reichstag, 9 September,1982.This volume.

Penkett,S.A.;Brice,K.A. and Eggleton,A.E.J.(1975)
A study of the rate of oxidation of sodium sulphite
solution by hydrogen peroxide and its importance to
the formation of sulphate in cloud-and rainwater.
Paper presented at Ispra(Italy),October 1975.
Technical Symposium,Project COST 61a.

Richards,L.W.;Anderson,J.A.;Blumenthal,D.L.;Mc Donald,J.A.
Kok,G.L. and Lazrus,A.L.(1982)
Hydrogen peroxide and sulfur(IV) in Los Angeles Cloud
Water.Paper presented at SCADDER conference,Santa
Monica,USA, 29 Nov.- 3 December,1982.

Ridder,T.B. and Frantzen,A.J.(1983)
Acid Precipitation over the Netherlands.Paper presented at Workshop on "Acid Deposition",Berlin,Reichstag
9 September1982. This volume.

Römer,F.G.;Viljeer,J.W.;Van den Beld,L.;Slangewal,H.J.,
Veldkamp,A.A. and Reijnders(1983)
Preliminary measurements from an aircraft into the
chemical composition of clouds.Paper presented at
Workshop on "Acid Deposition".Berlin,Reichstag,
9 September 1982. This volume.

Schwartz,S.E.;Daum,P.H.;Hjelmfeld and Newman,L.(1982)
Cloudwater Acidity Measurements and Formation Mecha-
nisms-Experimental Design.Paper presented at SCADDER
conference,Santa Monica,USA,29 Nov.- 3 Dec.1982.

Smith,R.A.(1872)
Air and Rain-The Beginnings of a Chemical Climatology,
London:Longmans,Green,and Co.,1872.600 pages.

Titoff,A.(1903)
Beiträge zur Kenntnis der negativen Katalyse in
homogenen Systemen.Zeitschrift für Physikalische
Chemie 45,pp.641-683,1903.

Ulrich,B.(1983)
Effects of acid deposition.Paper presented at Workshop on "Acid Deposition",Berlin,Reichstag, 9 Sept.
1982. This volume.

Waldman,J.M.;Munger,J.W.;Jakob,D.J. and Hoffmann,M.R.(1982)
Fog-water composition in southern California.Paper
presented at SCADDER conference,Santa Monica,USA,
29 Nov.- 3 Dec.1982.

Wallen,C.C.(1981)
    A preliminary evaluation of the WMO/UNEP precipita-
    tion chemistry data.MARC report No.22.

Winkler,P.(1983)
    Trend developments of precipitation pH in central
    Europe.Paper presented at Workshop on "Acid Depo-
    sition",Berlin,Reichstag, 9 September,1982.This volume.

Radke,L.F.(1982)
    Preliminary measurements of the size distribution of
    the cloud interstitial aerosol.Paper presented at
    SCADDER conference,Santa Monica,USA,29 Nov.- 3 Dec.,
    1982.

LIST OF PARTICIPANTS

ALLEGRINI, I.   Consiglio Nazionale delle Ricerche
Area della Ricerca di Roma
Via Salaria km 29.300 - C.P.10
00016 Monterotondo Stazione
I - ROMA

ANGELETTI, G.   Commission of the European Communities
Directorate General Science, Research
and Development
200,rue de la Loi
B - 1049 BRUXELLES

AUGUSTIN, H.   Commissariat aux Risques Naturels
25,Avenue Charles Floquet
F - 75007 PARIS

BAU, H.   Umweltbundesamt
Bismarckplatz 1
D - BERLIN 33

BAUER, A.   Technische Universität Berlin
Straße des 17.Juni 135
D - 1000 BERLIN 12

BECKER, K.H.   Universität-Gesamthochschule Wuppertal
Physikalische Chemie - FB 9
Gaußstraße 20
D - 5600 WUPPERTAL 1

BEILKE, S.   Umweltbundesamt
Feldbergstraße 45
D - 6 FRANKFURT

BERRESHEIM, H.   Universität Frankfurt
Referat für Umweltschutz
Robert-Mayer-Straße 11
D - 6 FRANKFURT

BIEHL, H.M.   KFA Jülich
Projektträger Umweltchemikalien
Postfach 1913
D - 5170 JÜLICH

BLOMMERS,A.H.   Geosens bv
Vliegveldweg 30
NL - 3045 NS ROTTERDAM(Zestienhoven)

BROSSET, C.   IVL
                              Swedish Water and Air Pollution
                              Research Institute
                              Sten Sturegatan 42
                              S - 402 24 GOTHENBURG

BRANDT, C.J.   VGB
                              Klinkestraße 27-31
                              D - 4300 ESSEN 1

BRUCKMANN, P.   Landesamt für Immissionsschutz(LIS)
                              Wallneyer Straße 6
                              D - 4300 ESSEN 1

COLACINO, M.   Istituto Fisica Atmosfera CNR
                              P. Luigi Sturzo 31
                              I - 00194 ROMA

COLIN, J.L.   Université Paris VII
                              Laboratoire de Chimie Minerale des
                              Milieux Naturels
                              2,Place Jussien
                              F - 75251 PARIS Cedex 05

COX, R.A.   Environmental and Medical Sciences
                              Division
                              AERE Harwell,Oxfordshire
                              OX11 ORA
                              GB - HARWELL

DELMAS, R.   Laboratoire de Glaciologie
                              2, Rue Très-Cloitres
                              F - 38031 GRENOBLE Cedex

DLUGI, R.   Laboratorium für Aerosolphysik und
                              Filtetechnik I
                              KFZ Karlsruhe
                              Postfach 3640
                              D - 7500 KARLSRUHE

ELSHOUT, A.J.   NV KEMA
                              Utrechtseweg 310
                              NL - 6800 ET ARNHEM

FENGER, J.   National Agency of Environmental
                              Protection
                              Air Pollution Laboratory
                              Risø National Laboratory
                              DK - 4000 ROSKILDE

FETT, W.   Bundesgesundheitsamt
                              Institut für Wasser-,Boden-u.Lufthygiene
                              Corrensplatz 1
                              D - 1000 BERLIN 33

FLYGER, H.   National Agency of Environmental
                              Protection
                              Air Pollution Laboratory
                              Risø National Laboratory
                              DK - 4000 ROSKILDE

FRANTZEN, A.J.   Koninklijk Nederlands Meteorologisch
                              Instituut
                              Postbus 201
                              NL - 3730 AE DE BILT

FRICKE, W.   Gesamtverband des deutschen
                              Steinkohlebergbaus
                              Friedrichstraße 1
                              D - 4300 ESSEN

FUGAŠ, M.   Institut for Medical Research and
                              Occupational Health
                              POB 291
                              Mośe Pijade 158
                              YU - 41001 ZAGREB

FUHRER, J.   Universität Bern
                              Botanische Institute
                              Pflanzenphysiologisches Institut
                              Altenbergrain 21
                              CH - 3013 BERN

GARBER, W.D.   Umweltbundesamt
                              Bismarckplatz 1
                              D - 1000 BERLIN 33

GEORGII, H.W.   Universität Frankfurt
                              Institut für Meteorologie und
                              Geophysik
                              Feldbergstraße 47
                              D - 6000 FRANKFURT

GOUTORBE, J.P.   Ministère des Transports
                              Direction de la Météorologie Nationale
                              Etablissement d' Etudes et Recherches
                              Météorologiques
                              73 - 77 Rue de Sevres
                              F - 92 BOULOGNE-BILLANCOURT

GRAVENHORST, G.   Laboratoire de Glaciologie
                              2,Rue Très-Cloitres
                              F - 38031 GRENOBLE Cedex

GREGOR, H.D.   Umweltbundesamt
                              Bismarckplatz 1
                              D - 1000 BERLIN 33

GUICHERIT, R.   Research Institute for Environmental
                              Hygiene(TNO)
                              NL - 2628 VK DELFT

GRAF HATZFELDT, H.   Schloß Schönstein
                        D - 5248 WISSEN/SIEG 2

HERBSLEB, H.   Bundesministerium für Ernährung,
                        Landwirtschaft und Forsten
                        Rochusstraße 1
                        D - 5300 BONN 1

HERTEL, W.   Umweltbundesamt
                        Bismarckplatz 1
                        D - 1000 BERLIN 33

HOV, Ø   Norwegian Institute for Air Research
                        P.O.Box 130
                        N - 2001 LILLESTRØM

ISERMANN, K.   BASF
                        Versuchsstation Limburgerhof
                        D - 6703 LIMBURGERHOF

ISRAEL, G.   Technische Universität Berlin
                        Sekr. KF 2
                        Straße des 17.Juni 135
                        D - 1000 BERLIN 12

JACOBSEN, I.   Deutscher Wetterdienst
                        Abteilung F
                        Frankfurter Straße 135
                        D - 6050 OFFENBACH

JAESCHKE, W.   Universität Frankfurt
                        Institut für Physikalische Chemie
                        Referat für Umweltschutz
                        Robert-Mayer-Straße 11
                        D - 6000 FRANKFURT

KALLEND, A.S.   Central Electricity Research
                        Laboratories (CERL)
                        Kelvin Avenue
                        Surrey KT22 7SE
                        GB - LEATHERHEAD

LACOMBE, R.   Umweltbundesamt
                        Bismarckplatz 1
                        D - 1000 BERLIN 33

LAHMANN,E.   Bundesgesundheitsamt
                        Thielallee 88-92
                        D - 1000 BERLIN 33

LENSCHOW, P.   Senator für Stadtentwicklung und
                        Umweltschutz
                        Lentzeallee 12-14
                        D - 1000 BERLIN 33

LESSMANN, E.   Kernforschungszentrum Karlsruhe
                      Weberstraße 5
                      D - 7500 KARLSRUHE

LÖBEL, J.   VDI-Kommission Reinhaltung der Luft
                      Graf-Recke-Straße 84
                      D - 4000 DÜSSELDORF 1

LORENZ, H.   Institut für Wasser-Boden-und Luft-
                      hygiene des Bundesgesundheitsamtes
                      Corrensplatz 1
                      D - 1000 BERLIN 33

MAMANE, Y.   Environm.Engineering Technicon
                      ISR - 32000 HAIFA

MARSH, A.R.W.   Central Electricity Research
                      Laboratories(CERL)
                      Kelvin Avenue
                      Surrey KT22 7SE
                      GB - LEATHERHEAD

MASNIERE, P.   Electricite de France
                      Direction des Etudes et Recherches
                      6,Quai Watier
                      F - 78400 CHATOU

MORELLI, J.   Université Paris VII
                      Laboratoire de Chimie Minerale des
                      Milieux Naturels
                      2,Place Jussien
                      F - 75251 PARIS Cedex 05

OTT, H.   Commission of the European Communities
                      Directorate General Science, Research
                      and Development
                      200,rue de la Loi
                      B - 1049 BRUXELLES

PAFFRATH, D.   Deutsche Forschungs-und Versuchsanstalt
                      für Luft-und Raumfahrt e.V.
                      DFVLR
                      Institut für Physik der Atmosphäre
                      Oberpfaffenhofen
                      D - 8031 WESSLING/OBB.

PANKRATH, J.   Umweltbundesamt
                      Bismarckplatz 1
                      D - 1000 BERLIN 33

PENKETT, S.A.   Environmental and Medical Sciences
                      Division
                      AERE Harwell,Oxfordshire
                      OX11 ORA
                      GB - HARWELL

PERSEKE, C.   Universität Frankfurt
                               Institut für Meteorologie und
                               Geophysik
                               Feldbergstraße 47
                               D - 6000 FRANKFURT

PERROS, P.   Laboratoire de Physico-Chimie de l'
                               Environnement
                               Université Paris Val-de-Marne
                               Av.du Général de Gaulle
                               F - 94010 CRETEIL Cedex

PINART, J.   Universite Paris VII
                               Laboratoire de Chimie Minerale des
                               Milieux Naturels
                               2,Place Jussien
                               F - 75251 PARIS Cedex 05

RONNEAU, C. J-M.   Université de Louvain
                               Laboratoire de Chimie Inorganique et
                               Nucléaire
                               Chemin du Cyclotron 2
                               B - 1348 LOUVAIN-LA-NEUVE

RÖMER, F.G.   NV KEMA
                               Utrechtseweg 310
                               NL - 6800 ET ARNHEM

RUDOLF, B.   Deutscher Wetterdienst
                               Frankfurter Straße 135
                               D - 605 OFFENBACH

RUDOLPH, J.   Institut für Atmosphärische Chemie
                               KFA Jülich, ICH 3
                               D - 5170 JÜLICH

SANDRONI, S.   Commission of the European Communities
                               Joint Research Centre,Ispra
                               Establishment
                               I - 21020 ISPRA(Varese)

SARTORIUS, R.   Umweltbundesamt
                               Bismarckplatz 1
                               D - 1000 BERLIN 33

SCHERER, B.   FU Berlin
                               Institut für Geophysikalische Wissen-
                               schaften
                               Fachrichtung Meteorologie
                               Thielallee 50
                               D - 1000 BERLIN 33

SCHMÖLLING, J.   Umweltbundesamt
                               Bismarckplatz 1
                               D - 1000 BERLIN 33

SCHULZE, R.   Bundesministerium für Wirtschaft
                     D - 5300 BONN

STANGL, H.   Commission of the European Communities
                     Joint Research Centre
                     Ispra Establishment
                     I - 21020 ISPRA(Varese)

STERN, R.   TU Berlin
                     Institut für Geophysiklaische
                     Wissenschaften
                     Fachrichtung Meteorologie
                     Thielallee 50
                     D - 1000 BERLIN 33

STIEF-TAUCH, H.P.   Commission of the European Communities
                     200, Rue de la Loi
                     B - 1049 BRUXELLES

TRENKLER, H.   Vereinigung Deutscher Elektrizitätswerke
                     (VDEW e.V.)
                     Stresemannallee 23
                     D - 6000 FRANKFURT 70

ULRICH, B.   Universität Göttingen
                     Institut für Bodenkunde und Wald-
                     ernährung
                     Büsgenweg 2
                     D - 3400 GÖTTINGEN

VERSINO, B.   Commission of the European Communities
                     Joint Research Centre
                     Ispra Establishment
                     I - 21020 ISPRA(Varese)

WARNECK, P.   Max-Planck-Institut für Chemie
                     Saarstraße 23
                     D - 6500 MAINZ

WENGENROTH, H.   TU Berlin
                     Sekr. KF 2
                     Straße des 17.Juni 135
                     D - 1000 BERLIN 12

WINKLER, P.   Deutscher Wetterdienst
                     Meteorologisches Observatorium
                     Frahmredder Straße 95
                     D - 2000 HAMBURG 65

ZELLNER, H.   Universität Göttingen
                     Institut für Physikalische Chemie
                     D - 3400 GÖTTINGEN

ZEPHORIS, M.   Observatoire de Magny les Hameaux
                     EERM
                     F - 78470 St. Remy les Chevreuse

ZIMMERMEYER, G.   Gesamtverband des deutschen
                           Steinkohlebergbaus
                           Friedrichstraße 1
                           D - 4300 ESSEN

INDEX OF AUTHORS